PRAISE FOR *FROM BACKWOODS TO BOARDROOMS*

From Backwoods to Boardrooms is a thoroughly researched and detailed account of the origin of a new investment asset class, which is now common in institutional investment portfolios and which substantially altered private timberland ownership patterns in the United States. It is a must-read for students of forest business and finance, as well as those interested in institutional timberland investments.

> CHARLES M. TARVER, founder and former president
> of Forest Investment Associates, Atlanta

Professor Zhang has written a remarkable book about the transformation of private forest ownership in the United States and around the world during the past forty years. The scale of the change—from lands once owned by integrated forest products companies to institutional investors—is truly revolutionary and is a story impacting tens of millions of acres of private landholdings and billions of dollars of investment. This revolution in ownership with its complex reasons and policy impacts has not been told with such reasoned and thoughtful attention before. This book is a must-read for anyone who contributed to this transition, were participants in the organizations who made it happen, or beneficiaries of its impact.

> PETER C. MERTZ, cofounder and former CEO of Global Forest Partners

This is the first book to provide a comprehensive assessment of the restructuring of commercial timberland ownership and management in the United States and around the globe. Dr. Zhang has provided a great service to the sector by capturing not only the technical and theoretical aspects of this episode but the human side as well. Interviews with the pioneers of modern timber finance help make this book engaging as well as informative and capture details of the history that could have otherwise been lost. It will be required reading for anyone interested in understanding the workings of the modern forest sector.

> DAVID N. WEAR, nonresident fellow, Resource for the Future

A seminal piece on institutional timberland investment.

> CLARK S. BINKLEY, former chief investment officer,
> Hancock Timber Resource Group and GreenWood Resources

T0133232

Daowei Zhang has written the seminal book on institutional investment in timberland, which will be the standard reference henceforth. The book has a comprehensive and insightful blend of economic theory about investments; a great history and rich stories about early and current institutional investors and individual leaders in the investment arena; analysis of the reasons for the trend of the huge rise in institutional investments in the last four decades; and comments on prospects for the future. This is a fascinating mix of scholarship, investment sector acumen, and enjoyable reading for all readers and investors in the forestry sector.

FREDERICK W. CUBBAGE, professor, North Carolina State University

Whether you are a student or a professional institutional investor, Professor Zhang's book provides a meaningful analysis of how tax policy drove profound changes in timberland ownership in the private and public sector over the last four decades.

MIKE COVEY, executive chairperson, PotlatchDeltic Corporation

This book is a goldmine of information on the history, economics, and investment vehicles in the forestry asset class.

DAVID BRAND, chief executive officer,
New Forests Asset Management Pty Limited

In this original and comprehensive research piece, Professor Zhang presents the history of institutional timberland investments and pioneers, analytical opportunities, current events and major TIMOs and REITs firms, financial performance of TIMOs and timberland REITs, regional and international timberland investments, and future promises. It is encyclopedic, and a must-read for anyone involved in forest investment and management in the United States or other countries.

F. CHRISTIAN ZINKHAN, founder and chairman of The Forestland Group

Professor Daowei Zhang has written an outstanding and thorough history and analysis of the timber investment history.

DUNCAN CAMPBELL, founder of Campbell Global

This is a novel and prime research work on commercial timberland investments in the United States as well as in the world. Everyone in the timberland investment field should have Dr. Zhang's book on their shelves. It also adds a significant value to the general discipline of forest economics.

BIN MEI, Hargreaves Professor of Forest Finance, University of Georgia

From Backwoods to Boardrooms
The Rise of Institutional Investment in Timberland

DAOWEI ZHANG

Oregon State University Press   Corvallis

The publication of this book was supported by a gift from Emmett F. Thompson, Dean Emeritus, Auburn University School of Forestry and Wildlife Sciences.

Library of Congress Cataloging-in-Publication Data
Names: Zhang, Daowei, author.
Title: From backwoods to boardrooms : the rise of institutional investment
 in timberland / Daowei Zhang.
Description: Corvallis : Oregon State University Press, [2021] | Includes
 bibliographical references and index.
Identifiers: LCCN 2021033624 | ISBN 9780870711428 (paperback) | ISBN
 9780870711435 (ebook)
Subjects: LCSH: Forest landowners. | Forests and forestry--Economic
 aspects. | Forest management. | Forest policy.
Classification: LCC SD387.L33 Z43 2021 | DDC 634.9/2--dc23
LC record available at https://lccn.loc.gov/2021033624

First published in 2021 by Oregon State University Press
Printed in the United States of America
Second printing 2022

Oregon State University
OSU Press

Oregon State University Press
121 The Valley Library
Corvallis OR 97331-4501
541-737-3166 • fax 541-737-3170
www.osupress.oregonstate.edu

To my family

Contents

Illustrations

FIGURES

TABLES

Foreword

CLARK S. BINKLEY

The first institutional investment in timberland apparently took place in 1288 when King Magnus III of Sweden granted the Bishop of Västerås a 12.5 percent interest in a mine and the surrounding land that centuries later became the foundational assets for Stora Enso. The Church of Sweden remains a major forestland owner and indeed was instrumental in the largest institutional timberland investment in Africa to date.

The modern era of institutional forestry investment began in the southern United States over six decades ago. Interest in the asset class started with insurance companies and then spread among different kinds of investors. Institutional forestry investors now include, along with insurance companies, pension plans, banks, endowments, foundations, and large family offices. Institutional forestry investment has expanded to all regions of the United States and to Oceania, Europe, Latin America, Asia, and a few parts of Africa—indeed, all continents except Antarctica (which apparently had trees long ago). Forestry investments have provided good risk-adjusted returns as well as many environmental, social, and governance (ESG) benefits, so this growth is not surprising. What is surprising is that despite this long history, total institutional investment globally likely does not exceed US$100 billion, a tiny amount in comparison with the US$4 trillion invested in private equity.

Daowei Zhang's seminal book *From Backwoods to Boardrooms: The Rise of Institutional Investment in Timberland* documents the evolution of this form of forest ownership over the past six-plus decades, from its earliest stages in the US South to its present global extent. Zhang has interviewed over one hundred participants in the industry in multiple countries, including some of the earliest ones. The book includes a thorough review of the literature, including not only readily accessible peer-reviewed publications but also unpublished PhD theses, industry magazines, business journals, and the extensive gray literature provided by timberland investment managers, investment consultants, and institutions themselves. All of this is written in a refreshingly informal style accessible to anyone interested in the subject, including anyone interested in forestry or sustainable investing more broadly.

Although the first institutional investments in timberland were in the form of debt, the book focuses mostly on equity investments. These equity investments include both private equity and publicly traded timber companies, now organized mostly as real estate investment trusts (REITs). Zhang appropriately focuses on the United States, where most of this investment takes place, but gives the global situation due attention. The content spans theoretical questions related to the rationale for including timberland in a mixed-asset portfolio all the way to such pragmatic matters as alternative ways to organize on-the-ground property management.

Three aspects will be of particular interest to this audience.

Through extensive oral histories, Zhang documents the early days of institutional timberland investment, starting in 1952 with a loan that Travelers Insurance Company made to St. Regis Paper to acquire 140,000 acres of cutover timberland on the Suwannee River in Florida. Other insurance companies, including ones that spawned timberland investment management organizations (TIMOs), soon followed. But Canal Industries, a wood dealer and landowner in South Carolina, beat them to the finish line with timberland investment partnerships with North Carolina National Bank, Eastern Airlines, and AT&T pension funds. Some of the current TIMOs talk about being "pioneers of the timberland investment business," but that title rests with these earlier financial adventurers, most importantly with Charlie Raper, who managed Travelers' forays in debt and equity investors and was Zhang's predecessor as the Peake Professor of Forest Economics and Policy at Auburn University.

Zhang also compares private equity timberland investments and timber REITs with respect to risk-adjusted returns as well as other dimensions of interest to investors such as alignment of interests, capacity to exploit market inefficiencies, and their conservation records. Some of these research results are new to the literature and will likely lead to new thinking about timberland investment and sustainable forest management more broadly in the United States and globally.

Zhang concludes with a most thoughtful commentary on the future of the timberland investment business. Of particular note are his thoughts on the continuing tension between public and private equity investments and on "forest-based natural climate solutions." The latter refers to the use of trees to sequester carbon dioxide from the atmosphere and store it first in trees themselves and then in long-lived wood products. It appears that forestry investments could provide some 25 to 35 percent of the net reductions needed to achieve Paris Agreement targets and limit climate change to less than two degrees Celsius. These investments would not only achieve an important environmental goal but would also offer institutions a far greater opportunity to invest in forestry than they now have.

Preface

One day in the spring of 1993, I went to the office of Dr. Clark S. Binkley, professor and dean of forestry at the University of British Columbia. I was then a PhD candidate, and Dr. Binkley was on my dissertation committee and was supervising me directly that year because my major professor, Dr. Peter H. Pearse, was on sabbatical. I asked him for possible recommendations for summer work that could use my academic background and skills. He told me to come back and see him in a couple of weeks.

Little did I know then that Dr. Binkley had just been asked to do a scientific review of the so-called Russell-NCREIF Timberland Index and thought that I could help. Russell stands for the Frank Russell Company, an investment services firm based in Tacoma, Washington, and NCREIF is the acronym for the National Council of Real Estate Investment Fiduciaries based in Chicago, Illinois. Apparently these two entities had been asked to construct a timberland index in the United States in 1992, which they did. The scientific review was to validate the methodology, construction, contributing guidelines, composition, and actual operations of the newly established index. Later, "Russell" was dropped from the name, and the index has since been called the NCREIF Timberland Index, as we know it today.

The request to create such an index came from a few institutional investors who had recently invested in timberlands during the preceding years. The index would serve as a benchmark for measuring the performance of their timberland investments. It was through helping review the index that I first learned about institutional timberland investment. Previously, nearly all the timberland owners in the United States that I had known about were public or government agencies, publicly traded and private industrial firms that also had forest products manufacturing facilities, and nonindustrial private forest (NIPF) owners that, by definition, do not own any forest products manufacturing facility.

This book is about the history and economics of institutional timberland investment, which started in the 1950s-1960s and has been on rise in the United States and elsewhere in the last four-plus decades. Concurrently, industrial timberland ownership, which collectively held some 14 percent of timberlands in the United States and accounted for nearly 30 percent of timber supply in the country in 1987, has almost disappeared. This is an unprecedented change.

The early 1980s were difficult for US forest products firms. They experienced a severe and prolonged economic recession. As a result, some went bankrupt, and a few were purchased by financiers who then sold them off in pieces. Some of them decided to treat their timberlands as separate profit centers, completely independent from their manufacturing business. Other firms decided to sell some of their timberlands outright or put them in master limited partnerships (MLPs) over which they still had some degree of control. The buyers of these timberlands in the mid- and late 1980s—rather big in size and value—were mostly institutional investors, including pension funds, insurance companies, banks, foundations, endowments, and high-net-worth individuals. The buyers of shares or units of timberland MLPs were individuals and some institutional investors.

Most institutional timberland investors have used investment management firms to purchase, manage, and sell their timberlands. These timberland investment firms have been referred to by some as timberland investment management organizations, or TIMOs. A long series of timberland transactions between industrial and institutional investors followed between the late 1980s and early 1990s, with a limited number of TIMOs advising the buyers. As institutional timberland investment increased and the TIMO business model gained acceptance, the number of TIMOs and assets under their management in the United States increased significantly.

In the late 1990s, timberland real estate investment trusts (timberland REITs) emerged and gradually replaced all existing timberland MLPs, and some previously publicly traded forest products companies were converted into REITs, such as Rayonier Inc. in 2004 and Weyerhaeuser Company in 2010. Together, these two forms of institutional timberland ownership—the purchase of shares of publicly traded timberland MLPs and REITs, and the direct investment in private equity via TIMOs or in house—grew by leaps and bounds in the United States, resulting in a nearly complete transformation by 2010 of the ownership of industrial timberlands.

Similarly, and perhaps because of the US experience, institutional timberland investment was on the rise in other parts of the world. It is this story—a rather large and long market event that has significant implications for the forest economy in the United States and the rest of the world—that I tell here. Further, I identify issues that have contributed to the rise and decline of industrial timberland ownership in the first place and explore how these events might be operative more universally in the realm of forest ownership and management. Finally, I look at the implications of institutional timberland investment and speculate on its future worldwide.

Over the last twenty-eight years, I have reviewed and studied the financial characteristics of timberland investment. I have also written articles that explain the logic and fallout of industrial timberland ownership. More importantly, I have been associated with and interviewed most of the pioneers and early practitioners of the TIMO business in the United States and elsewhere as well as the major players among institutional timberland investors in other countries. By putting all that I have learned about industrial and institutional timberland investment and ownership together in this book, I hope to shed some light on why institutional investors could profit more than industrial owners from the same lands and thus replace industrial owners in the United States, and how institutional timberland investment works in the rest of the world. I hope also to project the direction in which institutional timberland ownership is heading and to provide a better understanding of the effects of different ownership on forest management and sustainability.

Any success I may have achieved is shared with others. Auburn University has supported me on this and other research projects. Dr. Clark S. Binkley, who led me to encounter the subject of this study ahead of most of my peers, provided invaluable comments on the first draft and graciously wrote the foreword for this book. Many insiders—from industrial and institutional owners to TIMOs and REITs—gave their time for my interviews and provided their insights. The Market and Trade Program of the US Department of Agriculture National Research Initiative Competitive Grants Program (Grant No. 2011-67023-30051) provided funding for my earlier research on the subject, and Bradley/Murphy Forestry and Natural Resources Extension Trust in Birmingham, Alabama, provided financial support for further research that led to this book.

I would also like to express my appreciation to the following persons for their collective and individual assistance: to a few of my former graduate students including Xing Sun, Yanshu Li, and Noel Perceval Assogba for research assistance; to Duncan Campbell, Peter C. Mertz, Dale Morrison, Jake Petrosino, Scott C. Sacco, Sherry C. Smith, Charley Tarver, and the California Public Employees' Retirement System (CalPERS) for helping locate certain historical materials; and to Peter H. Pearse, William F. Hyde, and William Consoletti for their encouragement. Special thanks are due to Spencer B. Beebe, Craig Blair, Frederick W. Blum, William Bradley, C. Edward Broom, Thomas P. Broom, Amos Canal, Reid E. Carter, Jon Caulfield, Robert "Bob" G. Chambers, Douglas

Charles, Michael L. Clutter, Thomas J. Colgan, John Davis, L. Richard Doelling, John Earhart, Tracy Buran Evens, David Flowers, Bob Flynn, Bruno Fritschi, Barry Gamble, Eva Greger, Rick R. Holley, Jim Hourdequin, L. Scott Jones, L. Michael Kelly, Worth Kendall, Jack Lutz, Dick Molpus, Jeff Nuss, Hank Page, Cotty Peabody, James "Jim" A. Rinehart, Kate Robie, David M. Roby, Richard N. Smith, Peter R. Stein, Charles L. VanOver, Marc A. Walley, Courtland L. Washburn, Jim Webb Jr., Ben Whitaker, Paul Young, and F. Christian Zinkhan for their insights and perspectives. I also acknowledge insights generated from conversations with Mads Asprem, Olli Haltia, Marko Katila, Christophe Lebrun, Petri Lehtonen, Pedro Ochoa, Tapani Pahkasalo, and Nils von Schmidt on European and African timberland investments, and with Bryce Heard and Dennis Neilson on Oceanian timberland investments. Unfortunately, I may miss some names here, for which I apologize. A few of these insiders saw the relevant chapters of this book in their penultimate format and provided further insights and comments, for which I am indebted to them, but none of them completely agreed with my perspective nor did they endorse any part of these chapters. Three anonymous referees and the editor at Oregon State University Press provided invaluable comments.

Finally, I would like to acknowledge my appreciation for my family. My wife, Zilun Fan, has supported my effort throughout this project and many other works. Our two kids, Xinrei and Ting Dan, are vivid critics and helpers of their dad's work.

CHAPTER 1
From Industrial to Institutional Timberland Ownership

For more than a century prior to 1990, forest products companies in the United States had been accumulating timberlands, mainly to provide wood for their mills. Because these companies are industrial firms, they are called industrial timberland owners. These companies are also called vertically integrated forest products companies because they integrate their forest products manufacturing plants with timberlands. At the peak in 1987, total industrial timberlands in the United States reached about seventy-one million acres and supplied nearly 30 percent of the timber produced in the country.[1] Although forest products firms started to sell some pieces of timberland in the mid-1980s, few at that time could foresee that the demise of traditional industrial timberland ownership was imminent and that all publicly traded forest products companies would eventually sell their timberlands or concentrate on timberland business by selling most of their forest products manufacturing facilities in the next two decades. In retrospect, this move was seen as the beginning of the end of industrial timberland ownership in the United States.

Indeed, most large forest products companies, such as Boise Cascade, Bowater, International Paper, Kimberly-Clark, Louisiana-Pacific, Smurfit-Stone, Temple-Inland, MeadWestvaco, and Weyerhaeuser Company all started to sell their timberlands in large quantities in the late 1980s and 1990s. Georgia-Pacific Corporation, one of the largest forest products companies and one of the three largest industrial timberland owners in the United States during that time, was the first to decouple nearly all of its six million acres of timberland from its manufacturing business. It first set up a separate timberland entity called The Timber Company in 1997 and then sold it to Plum Creek Timber Company in 2001.

The selling of industrial timberlands intensified between 2000 and 2006. Eventually, International Paper Company, the largest timberland owner in the United States by the early 2000s, announced on July 19, 2005, that it would sell most of its timberlands—some 6.8 million acres in the United States—which was mostly completed in April 2006.[2]

The Weyerhaeuser Company, the last of the three largest industrial timberland owners, converted itself to a timberland real estate investment trust (REIT) in 2010 by selling some of its forest products manufacturing facilities, by packaging other manufacturing facilities as a non-REIT subsidiary, and by concentrating on its timberland business. By the end of 2010, the divestment of industrial timberland ownership had largely been completed in the United States. Altogether, some fifty million–plus acres of industrial timberlands were sold or transferred to different ownership. A bit of these industrial timberlands was sold in small pieces to private individuals. Most were purchased by institutional investors, whose investments were often held in commingled funds or separate accounts managed by private investment management firms, known by many as timberland investment management organizations (TIMOs).[3] Some of these industry-owned timberlands remained in public markets but were transferred into publicly traded timberland REITs. The list of the largest timberland owners in the United States, dominated by forest products companies thirty years earlier, was completely different in 2011 (fig. 1.1).

Why did industrial firms want to own timberlands in the first place? Relatedly, what made them want to sell the timberlands they had accumulated over many decades, mostly to institutional investors, in the twenty years between 1990 and 2010? Who are these institutional investors? Why would institutional investors be interested in purchasing timberlands? How were institutional investors able to buy timberlands from industrial owners, and what were the conduits of these transactions? What additional value could institutional investors gain from owning timberlands that industrial owners could not?

From a broad perspective, institutional investors are large investment entities that pool large sums of money and invest those sums in securities, real property, and other investment assets. Institutional investors include banks, insurance companies, retirement or pension funds, endowments, foundations, hedge funds, and mutual funds as well as offices of ultrahigh-net-worth individuals and families—all hold investment assets as fiduciaries for the benefit of others. Timberland investment is investment in timber and timberland through direct ownership and long-term leases for the purpose of producing attractive, risk-adjusted returns during the period of investment.

All institutional investments—in timberlands or anything else—are for the purpose of generating financial returns and meeting other investment needs. Institutional investors invest in timberlands mainly in three ways: through professional investment advisers and timberland managers

Figure 1.1. Structural shift of industrial timberlands to institutional ownership, 1981, 1994, and 2011

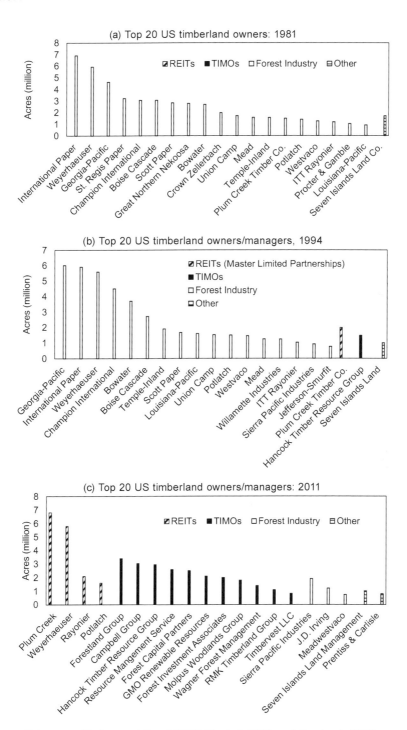

Data source: Clephane and Carroll (1982), Yin et al. (2000), and Hourdequin (2013).

(TIMOs); through companies that own and control publicly traded timberland real estate investment trusts (timberland REITs); or through direct (inhouse) timberland investments. In the first option, institutional investors hire TIMOs to search for, purchase, manage, generate income from, and sell their timberlands, and their investments are considered private equity. As of 2020, twenty million of the nearly forty million acres of institutional timberlands in the United States were under the management of TIMOs.[4] US-based TIMOs also managed/owned 9 million acres of timberland elsewhere in 2020.

In the second option, institutional investors own shares of publicly traded timberland REITs, just as they would publicly traded stocks through mutual funds and hedge funds. As of 2020, there were four publicly traded timberland REITs in the United States: Weyerhaeuser Company, Rayonier Inc., PotlatchDeltic Corporation, and CatchMark Timber Trust Inc.; and one in Canada (Acadian Timber Corp.), which owned some timberlands in the United States. Acadian Timber Corp. is listed on the Toronto Stock Exchange, although institutional investors can trade its shares in OTC markets in the United States as well.

Timberland master limited partnerships (timberland MLPs), a predecessor to timberland REITs, were also an investment vehicle for institutional investors. One such publicly traded timberland MLP—Pope Resources—lasted for thirty-six years until it was purchased by Rayonier Inc. in January 2020. Pope Resources and other timberland MLPs are much like timberland REITs and are treated as public equity timberland investments in this book. At the end of 2020, these timberland REITs collectively owned more than 16.6 million acres of timberlands in the United States and nearly 1 million acres of additional timberlands in other countries. And more than 80 percent of the common shares of these timberland REITs were owned by institutional owners (table 1.1). The remainder were owned by retail investors and insiders.

A few institutional investors directly own and manage timberlands in house. Although there are no data on the acreage under direct institutional timberland investment, it is believed to be much smaller than the total acreage either managed by all TIMOs or owned by all timberland REITs and perhaps accounted for less than 5 percent of all institutional timberland ownership in the United States at the end of 2020. Thus, institutional timberland investment in the United States is mainly through TIMOs and timberland REITs, even though this book covers all three forms of institutional investors.

This book is a study of the history, economics, finance, contemporary development, management, and future of institutional timberland

Table 1.1. Publicly traded timberland REITs in the US: Timberland acreage and shares owned by institutional investors as of December 31, 2016, and December 31, 2020

Name[a]	Timberland owned in the US (1,000 acres)	Timberland owned in other countries (1,000 acres)	Common shares owned by institutional investors (%)
2016			
Weyerhaeuser Company	12,800[b]	300[c]	75.4
Rayonier Inc.	2,300	435[d]	82.5
Potlatch Corporation	1,400[e]		83.2
CatchMark Timber Trust Inc.	500[f]		70.2
Pope Resources	131[g]		15.3[h]
Acadian Timber Corp.	299	761[h]	
Total	**17,434**	**735**	
2020			
Weyerhaeuser Company	10,700[b]		81.3
Rayonier Inc.	2,273[i]	417[d]	85.5
PotlatchDeltic Corporation	1,800		82.5
CatchMark Timber Trust Inc.	1,508[f]		81.5
Acadian Timber Corp.	300	761[h]	
Total	**16,581**	**417**	

a Data for timberland ownership are from the 2016 and 2020 annual reports of these respective companies; data on common stock shares owned by institutional investors are from Fidelity Investments, accessed January 1, 2017, and January 1, 2021, respectively, www.fidelity.com.

b Including 1.1 and 0.8 million acres of leased timberland in 2016 and 2020, respectively, and excluding timberland leased in Canada.

c In Uruguay, including 10,000 acres of leased timberland.

d In New Zealand.

e Including 18,000 acres of leased timberland.

f Including 468,000 acres and 408,000 acres of wholly owned lands in 2016 and 2020, respectively, and 32,000 acres of leased lands and 1.1 million acres of timberland under joint venture in 2016 and 2020, respectively.

g Consisting of 118,000 acres of timberland, 10% of coinvestment in 8,800 acres of owned Timber Funds, and 2,200 acres of development property. Pope Resources is a publicly traded master limited partnership (MLP) and had a market capitalization of only $290 million at the end of 2016. In addition, its insider ownership was 19.9%.

h In Canada, excluding 1.3 million acres of Crown lands under management.
i Including 17,000 acres owned in the private equity timber fund business under Olympic Resource Management.

ownership in the United States and in other parts of the world. My primary focus is the shift from industrial to institutional timberland ownership in the United States, where the modern version of institutional timberland investment started and where the size and extent of institutional timberland ownership is the largest in the world. I will also cover institutional timberland investment elsewhere because such investment has implications for sustainable forest management worldwide. Because industrial timberland ownership preceded institutional timberland ownership, I must explain the rise and fall of industrial timberlands, and the associated transformation of US and global forestry investing.

In particular, I investigate

(1) the fundamental values of timberland under perfect and imperfect market conditions that may lead to profitable maneuvers through different ownership and management actions;

(2) the historical roots and logic of industrial timberland ownership and the entrepreneurship, changing market conditions, and policy factors that have given rise to institutional timberland ownership as the new and increasingly important class of timberland ownership;

(3) the management and land use behavior of institutional timberland owners as well as the financial performance of TIMOs and timberland REITs;

(4) the broad implications of institutional timberland ownership for forestry investing, forest sustainability, and the conduct of the whole forest sector in the United States and elsewhere; and

(5) the potential future of institutional timberland investment in the United States and the rest of the world.

This large shift in forestland ownership in the United States is interesting in its own right, as it was "unprecedented and represented the single largest private landownership change in the nation's history."[5] Ownership objectives are known to differ substantially between industrial and nonindustrial private owners.[6] The latter include both individuals and corporations who do not own the equipment to convert logs into lumber, pulp, or other forest products. The ownership objectives of institutional timberland investors would certainly be different from those of industrial owners and traditional nonindustrial private owners. Further, and perhaps more importantly, the ownership of timberlands determines the economic efficiency of forest resource use and the distribution of income associated with the utilization of these resources. Will timber supply, forest regeneration, and forest management and protection change under institutional owners in the United States and elsewhere? Will forests be fragmented, with impacts on recreation and biodiversity? Will this change in ownership affect forest workers and communities that depend on the forest industry, and the forest research and development that are drivers of forest growth in the United States and other parts of the world in the twenty-first century? What will forestry investing in the United States and globally look like in the future?

The search for an economic explanation for such a shift is also interesting, as it is a significant part of the economic history of the forest sector in the United States and beyond. This search may provide insights into the shift in corporate management paradigms from manager control of conglomerates to shareholder value, the market dynamics of industrial timberlands before the 1990s and afterward, transaction cost economics, and institutional and policy factors such as corporate income tax and dividend tax policy, accounting rules, and corporate strategies. As we will see, all these things have contributed to the imperfections observed in timber and timberland markets and thus have given rise to opportunities for arbitrage—first with the emergence of profitable industrial timberland ownership, and then with a shift from industrial to institutional timberland ownership. For example, individual timberland owners and corporate timberland owners are known to have different rates for income taxes, and different corporate structures (C corporations vs. S corporations and limited liability corporations) imply different tax liabilities. Finding out how each of these factors contributes to the rise and fall of industrial timberland ownership would be of interest to policy makers, managers of forest products firms, forest workers, forest-dependent communities, and anyone interested in forest management and sustainability.

Finally, TIMOs of a sizable scale have had a history of only about forty years, and TIMOs and timberland REITs have coexisted for only a little more than twenty years, or nearly forty years if timberland MLPs are considered. Which one better aligns with the interests of investors and management? Which one has performed better financially? Is there room for both forms of institutional timberland investment to grow? Without doubt, competition and mismanagement have put some TIMOs and REITs out of business and will continue to do so. Because the conversion of industrial timberlands to institutional ownership is largely completed, how will they adapt to changing market conditions, develop new strategies, and foster future growth? Are they both starting to buy nonindustrial private forestlands and to sell to each other simultaneously? What are the implications for forest sustainability as institutional investments increasingly go global?

I would like to address all these questions. I will start with industrial timberland ownership because that is where institutional investors acquired most of their timberlands in the United States. The central hypotheses of this book are (1) that industrial timberland ownership rose as industrial entrepreneurs explored inefficiencies in timber and timberland markets and sought sustainable supplies of timber for their manufacturing operations; (2) that this ownership declined because those market inefficiencies were largely eliminated by developments in transportation, communication, and especially tree-growing technology that transformed timber and timberland market economics; and (3) that from the mid-1980s onward, institutional investors have been able to exploit other market inefficiencies that were detrimental to industrial timberland ownership, including valuation metrics, transaction cost economics, tax policy, and accounting rules. Also, timber and timberland market efficiency improves as more institutional investors enter the markets and push investment frontiers further to seek profitable opportunities. Thus, after the idiosyncratic risk associated with individual investments is considered and incorporated, the risk-adjusted returns from all timberland investments should in the long run reflect differences in forest type, tree species, growth rates, and location, including differing country risks.

The remainder of this chapter provides a brief history of industrial timberland ownership in the United States, reviews the economic rationale and desire of industrial firms to secure timber supply from their own lands and from nonindustrial forestlands, and presents the objectives, research methods, and organization of this book.

A Brief History of Industrial Timberland Ownership

The forest products industry consists of three industry groups: wood products, paper and allied industries, and wood furniture. In the nineteenth century, the industry was dominated by lumber producers who, in order to access mature timber, bought large amounts of timberland in the country, especially in the East and South.[7] However, forest products firms at that time were operating in an extractive mode, as evidenced by portable sawmills that moved from New England to the Lake States to the South and finally to the Pacific Northwest to access mature timber. Further, paper mills were in their infancy and were concentrated mostly in the Northeast and to a lesser extent the Midwest. Because mature timber was plentiful in the United States at that time, firms could easily move to other locations if mature timber in one location was depleted, thus leading to portable "peckerwood" sawmills. Instead of holding on to timberlands and waiting for second-growth forests to mature, forest products firms often sold cutover tracts for farmland or even abandoned them and let them revert to counties for nonpayment of taxes.[8]

Thus, until early in the twentieth century, forest products firms, which were rather small, showed little interest in timber management and long-term, large-scale ownership of timberlands. Their main motive for owning timberlands was to access the mature timber on the land. Once the mature timber was harvested, they often did not want to keep the land, as their strategy then was simply to "cut out and get out."

This all changed as the size of sawmills and especially pulp and paper mills grew and the amount of timber that these mills required at a given location increased drastically in the early twentieth century. Once set up, these new mills were not going anywhere. As a result, the mills depended for their timber on what could be grown in the immediate procurement area of about a 150-mile radius. The earlier extractive logic employed by the lumber industry no longer sufficed. It was around this time that forest products companies began to accumulate lands for the purpose of growing trees, which would in turn protect their large capital investments in their mills. Let us use International Paper Company and Weyerhaeuser Company to illustrate the concentration of timberlands in the US forest industry for much of the twentieth century.

* * *

International Paper Company was incorporated in Albany, New York, on January 31, 1898. International Paper became a firm as the result of a merger of seventeen pulp and paper mills. Its initial production facilities ranged from a small mill that produced 11 tons of paper per day to one of the most advanced in the industry with a daily output of 150 tons. Its initial holdings also included 1.7 million acres of timberland in the northeastern states and Canada. In its early years, International Paper was the nation's largest producer of newsprint, supplying 60 percent of all newsprint sold in the United States and exporting to Argentina, England, and Australia.[9] As the company's newsprint business in the United States declined after 1913 when the US Congress eliminated tariffs on low-cost Canadian imports, it formed a subsidiary in Canada called Canadian International Paper that acquired vast tracts of timberland in Canada in fee-simple ownerships and mostly timber leases from Crown (government-owned) lands. It also constructed some of the world's largest newsprint mills in the province of Quebec.

In its 1916 annual report, the president of International Paper, P. T. Dodge, made this statement:

The Company is also the owner of very extensive woodlands, both in the United States and Canada, and is also the possessor of extensive Canadian Crown lands, carrying the right to cut wood in Canada. The woodlands are of great and increasing value, and in the years to come will find a great competition in assuring a supply of pulpwood to the mills of the Company. The available woodlands of the country are steadily diminishing and the value of wood increasing. *It is believed that the timberland holdings of the Company will protect its mills. In the years to come, many competitors will be compelled to obtain wood, if it is to be had at all, at almost prohibitive prices*[10] (emphasis added).

International Paper did not reveal the size of its timberland ownership in its annual report again until 1934, when it reported it owned about 2 million acres. Nonetheless, it is safe to assume that its total timberland acreage in the United States was at least 1.7 million acres in 1919 because the president of the company stated that it had a policy of increasing its woodland holdings.[11] However, the book value of these timberland assets had been down from $5 million in 1898 to $3 million in 1916 and $2.1 million in 1920, similar to the Weyerhaeuser Company's conversion from timber assets to manufacturing assets (see below).[12] Subsequently, International Paper's timber procurement policy was "to purchase wood from outside sources, conserving its own timber wherever the age and condition are such that the growth will meet the carrying charges."[13]

In its 1920 annual report, the president of International Paper reiterated the danger of wood shortage and rising cost, the absolute necessity of speedily acquiring additional woodlands, purchasing outside wood for the time being, and cutting its own wood in a conservative manner. He stated:

The ideal condition would be the ownership of land in such quantity that under judicious cutting the growth would equal the cut, and thus provide a perpetual supply for your mills. This cannot be accomplished without the investment of a great amount of additional capital, probably unwise and unobtainable on reasonable terms at the present day.

The fact that this capital cannot be advantageously obtained and that the purchases cannot be delayed, *make it necessary that the current earnings or profits of the Company shall be largely invested for a considerable period in wood protection*[14] (emphasis added).

When International Paper diversified into southern kraft paper production by acquiring the Bastrop Pulp and Paper Company in Bastrop, Louisiana, in 1925, it immediately took "an option on a tract of woodland which, if investigations now under way prove up, will assure the enlarged mill at reasonable cost of *a permanent supply of raw materials from its own lands*"[15] (emphasis added).

International Paper then stepped up its asset acquisition in the South. It purchased the Louisiana Pulp and Paper Company in 1927 and another mill in Moss Point, Mississippi, shortly afterward. It then created a new subsidiary, the Southern Kraft Division (initially called Southern International Paper), to manage its southern operations. Between 1929 and 1931, it built three new paper mills: one in in Camden, Arkansas; one in Mobile, Alabama; and another in Panama City, Florida. This division was the leading pulp producer in the southern United States by 1930, earning profits even during the depths of the Depression and making up for International Paper's unprofitable northern operation.[16]

The large investments in fixed capital embodied in a new pulp and paper mill and the substantial demand for a continued supply of timber to feed the mill led International Paper and other paper makers to integrate backward into raw materials supply. The capital investments were so great that it would

Figure 1.2. Timberland owned by International Paper Company in the US, 1898–2008

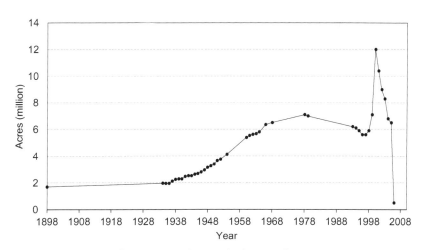

Source: International Paper Company annual reports (various years).
Note: International Paper reported its timberland acreage in its annual report only in 1898, 1934, and some years afterward. The peak of timberland ownership around 2000 occurred because the company bought Union Camp in 1999 and Champion International in 2000.

have been reckless not to invest in a sustainable supply of timber. And at the time, forest products companies could do so at extraordinarily low costs. Some timberlands were even abandoned after timber harvesting to avoid paying property taxes, so it could be argued that International Paper often set the market prices for rural lands because it was the only large acquirer in a given area.[17]

In 1948, International Paper owned 3.2 million acres of timberland, of which some 2 million acres were in the US South, and its production and revenue from mills as well as the acreage of timberland in the South surpassed that in the Northeast (fig. 1.2).[18] As International Paper grew its production capacity, the size of its timberland holdings increased steadily. The spike in its timberland acreage around 2000—some 12 million acres—was the result of its purchase of Union Camp Corporation in 1999 and Champion International Corporation in 2000, which together had owned nearly 6 million acres of timberland. As noted earlier, International Paper gradually sold about half of its timberland between 2000 and 2005, and the other half in 2006.

* * *

In 1900, Frederick Weyerhaeuser and fifteen partners founded the Weyerhaeuser Timber Company (now Weyerhaeuser Company) in Tacoma, Washington, by purchasing 900,000 acres of timberland with mature timber from the Northern Pacific Railway Company in the state of Washington at the price of six dollars per acre.[19] This was the largest private land transaction in American history at that time.[20] By 1920, Weyerhaeuser had about 2 million acres of timberland, all in the Pacific Northwest. In the next two decades, the total timberland acreage owned by Weyerhaeuser changed little, and in 1945, the company had 2.03 million acres.[21]

While accumulating timberlands in the first two decades of the twentieth century, Weyerhaeuser always wanted to generate value by converting timber to lumber and other forest products. Thus, in 1903, it opened its first sawmill in Everett, Washington. In 1929, it built its second sawmill in Longview, Washington—the largest sawmill at that time—and a pulp mill at the same site two years later. The paper mill allowed the company to utilize its vast holdings of hemlock, and it turned out to be a "Great Depression miracle," making a profit while overall the company sustained losses. Between 1945 and 1949, Weyerhaeuser opened three kraft pulp mills; a liquid packaging board mill; plywood and ply-veneer plants; particleboard, containerboard, hardboard, and wood-fiber mills; and even a bark-processing plant.[22]

The conversion of timberland assets to other capital assets such as plants and machinery in Weyerhaeuser was evident in its 1949 annual report:

In 1915, net assets of $159.5 million were largely static mature timberlands: only 5 percent of the area owned had been logged. By the close of 1945 the timber harvest from 43 percent of the area owned had provided plants and other working assets for the company and its subsidiaries, leaving less than one-half its assets in mature timber.[23]

The second wave of timberland accumulation by Weyerhaeuser occurred after the Second World War, especially after it decided to expand to the South. In 1956, it first purchased 90,000 acres of timberland in Mississippi and Alabama. In 1957, it added 460,000 acres of timberland in North Carolina, Virginia, and Maryland, and by then, it had manufacturing facilities in nineteen states, bringing the company into the packaging, milk-carton, and folding-box businesses. When it became a public company and listed its stock on the New York Stock Exchange and the Pacific Coast Stock Exchange in 1963, Weyerhaeuser had 3.6 million acres of timberland across the United States, of which 2.9 million acres were in the Pacific Northwest, 712,000 acres in the South, and another 28,000 acres under long-term lease in the South.[24]

Weyerhaeuser's largest timberland purchase in the South occurred in 1969 when it bought Dierks Forests Inc. for $325 million, adding 1.8 million

Figure 1.3. Timberland owned by Weyerhaeuser Company in the US, 1900–2010

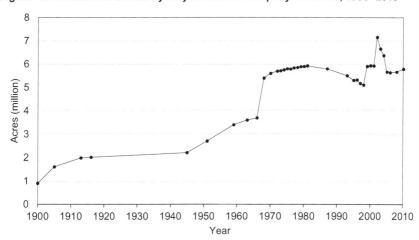

Source: Weyerhaeuser Company annual reports, various years.

acres of timberland in Arkansas and Oklahoma to its previously owned lands and thereby bringing its total US timberlands to 5.4 million acres. Since then, Weyerhaeuser's timberland acreage in the South has surpassed that in the Pacific Northwest. Its total timberland ownership in the United States stayed around 5–6 million acres for the next thirty years until it bought Willamette Industries Inc., which had 1.7 million acres of timberland in 2002 (fig. 1.3). However, in order to pay down its debt, which occurred in financing its purchase of Willamette Industries Inc., Weyerhaeuser sold some 2 million acres of timberland and other manufacturing assets, bringing its timberland holdings to about 6 million acres by 2005. As noted, Weyerhaeuser converted from an integrated forest products company, a C corporation, to a timberland REIT as an S corporation in 2010.

The Benefits and Costs of Industrial Timberland Ownership

In the context of industrial organization theory, the ownership of timberlands by a forest products company is upstream or backward vertical integration: integration using timber as an input in the production of forest products. Broadly, the rationale and benefits of upstream vertical integration are supply assurance, imperfect competition, barriers to entry, and transaction cost economies (that is, low transaction costs for timber from a forest industry firm's perspective), among others.[25] Strictly, all these reasons could be summarized as imperfect markets, as some conditions for a perfect market fail in economic terms. Usually, the conditions for perfect competition or a perfectly competitive market include

- a large number of buyers and sellers, all being price takers (no participant can have market power to set prices);
- perfect information: all buyers and sellers know the prices of products and inputs and the profits/utilities they would get from each product/input;
- homogeneous products: the products are perfect substitutes for each other and do not vary among different suppliers;
- well-defined property rights;
- no barriers to entry or exit;
- perfect factor mobility, which allows free adjustments in all factors of production to changing market conditions;
- profit maximization of sellers and rational buyers;
- no externalities: costs or benefits of an activity do not affect any third parties;

- zero or very low transaction costs: buyers and sellers do not incur high transaction costs in making an exchange of goods and services; and
- no increasing returns to scale and network effects: the lack of economies of scale or network effects ensures that there will always be a good number of firms.

In the long run, economic profit does not exist in perfectly competitive markets. Should one or more of these conditions fail, there would be firms to exploit profitable opportunities. As we will see later, timber and timberland markets in the United States and elsewhere are far from perfect competition, giving forest products companies an opportunity to own timberlands and generate economic profit consistently for a long time.

Let us examine the benefits or advantages of industrial timberland ownership noted in the literature, including favorable returns on timberland, supply assurance, cash flow and earnings stabilization, cost control, leverage on open stumpage markets, tax advantages (prior to the 1986 tax reform in the United States), and reduction of price volatility and risk.[26]

Favorable returns on timberlands are relative. Over time, forest products companies may generate returns from their timberlands that are comparable to or higher or lower than their own cost of capital. When timberlands as a stand-alone asset do not generate adequate returns that are comparable to the cost of capital, the question arises: Why own these timberlands? This was the question James A. Rinehart asked in 1984. After examining the cost of capital associated with industrial timberlands, Rinehart argued that forest products firms might just be better off raising capital by selling their timberlands rather than selling equity or debt.[27] The logic of favorable returns may also apply to a situation in which a forest products firm is able to lower its overall cost of timber and thus increase its profit.

Supply assurance or wood security is the most often cited reason for industrial timberland ownership. Timber as a raw material is a significant element in the total cost of producing wood and paper products. Using the southern United States as an example, Morgan Stanley estimated that timber accounted for 60–70 percent of the total manufacturing cost of lumber and plywood and 20–30 percent of that of white paper and linerboard in the early 1980s.[28] As mill sizes increase and companies operate under an imperative to keep their fixed capital in production, many paper mills run twenty-four hours a day, seven days a week and thus demand high volumes of timber, water, and energy. Because harvested timber tends to decay quickly, storage as a form of supply assurance can work for only days or weeks and is thus not a good option. As a result, these mills depend on the availability of timber that can be grown in their immediate procurement areas. Thus, when companies decide to build a new wood products or paper mill, they often look for places that have adequate timber resources and purchase some timberlands for supply assurance. Simply put, these mills need wood, but not land. Owning timberland is just a means of ensuring timber supply.

But self-sufficiency or self-supply is not necessary if adequate timber is readily available at a reasonable price in open markets within the timber procurement area of a manufacturing plant. When the timber market is competitive, supply assurance comes into play only if bad weather or an unexpected event causes other landowners to not want to harvest timber on their lands because of environmental concerns such as soil compression. Therefore, the amount of timberland needed for supply assurance for bad weather is relatively small.

Earnings stabilization and risk reduction imply that timberland and forest products manufacturing businesses have different and somewhat unrelated financial characteristics in terms of returns and risk. Thus, companies could manage the cash flows from these two segments of business to stabilize their overall cash flow and earnings and reduce their inherent business risk.[29]

For instance, when forest products markets are strong, strong timber markets often follow. Companies can buy more timber from open markets at relatively high cost and thus dampen or stabilize their cash flow and earnings in good market conditions. On the other hand, when forest products markets are weak and timber markets are weak, they can use more of their own timber, albeit at lower prices, thereby boosting their cash flow and earnings in bad market conditions.

This cash flow and earnings stabilization strategy often allows forest products companies to smooth their cash flow and earnings and subsequently reduce the volatility of their stock prices over time, which can protect corporate executives' jobs. However, this corporate strategy—chasing low risk at corporate levels via upstream integration and internalization of timber operations—does not necessarily maximize or align with shareholder value. It is thus not what the shareholders of the companies prefer or want, especially when a significant segment of shareholders—the managers of mutual funds that own the stocks of these companies—begin to question the wisdom of this practice, because investment diversification can be achieved at the shareholder level.

This leaves us with cost control, hedge (or leverage) on open stumpage markets, and tax advantages (before 1986), which are all about minimizing the cost of raw materials. Collectively, these reasons may be called the leverage theory of industrial timberland ownership.

The bulky nature of timber implies that logs cannot be transported for a long distance and that an imperfect timber market condition often exists. Industrial timberland owners could reduce the cost of the timber they purchase from open markets by using timber produced from their own lands as leverage. This is to say that owning timberlands may increase the economic profit of a forest products firm consistently in the long run.

As the appendix to this chapter shows, if a firm is the sole or dominant buyer of timber in its procurement region, it can generate monopsonistic rents.[30] Obviously, the example used in the appendix is highly simplified, and alleged exertion of monopsonistic powers on timber suppliers could lead to antitrust lawsuits, as International Paper Company experienced in South Carolina in the early 2000s.[31] Furthermore, increasing competition would increase market efficiency and reduce or completely eliminate monopsonistic rents.

Nonetheless, empirical evidence shows that imperfect competition exists in US pulpwood markets,[32] and at a minimum, owning some timberlands may allow forest products companies to reduce the transaction costs of timber procurement.

Relatedly, industrial timberland owners may also create barriers to entry, fencing off potential competitors from entering the same market. Indeed, if the initial motive for forest products mills to acquire land bases was primarily to ensure a supply to support their multimillion-dollar mill investment, this move had the added benefits of price stability when competition for timber intensified, and possible monopsonistic rent in timber markets if competitors were prevented from entering the markets. As William Boyd states: "Owning timberland provided a hedge against high prices."[33]

In short, under imperfect competition and preferential tax treatment, industrial timberland ownership may enhance a firm's profitability by increasing its revenues from its own timber production and/or reduce its costs by controlling stumpage prices through timberland ownership or lease. But it needs only timber, not land. And this possible enhancement in profitability must be weighed against the capital costs of owning timberlands.

The preferential capital gains tax treatment for timber ceased to exist for a number of years after the 1986 tax reform in the United States. Additionally, because forest products firms are C corporations, their profits are "double taxed," first at the corporate income level and second when shareholders receive dividends from these corporations. In contrast, other entities—such as limited partnerships or REITs—are taxed only when their owners receive distributions. Furthermore, growing timber is very capital intensive, and a large amount of capital can be tied to timberlands.[34]

Finally, since timberlands and forest products production are two different businesses, conflicts and internal organizational costs may arise.[35] For example, when companies decide to use their timberlands to supply their mills rather than manage them as a profit center, they may underprice their own timber.[36] I have heard stories of companies' sawmill managers who offered lower stumpage prices to their timberland managers because these prices were all they could pay to make their sawmills profitable. This practice not only lowers the returns on timberlands but also causes potential conflicts and raises internal organizational costs, especially when industrial timberland holdings are large. It also shows the challenge faced by management within forest products firms in balancing the long-term fundamental nature of the business of growing trees with the immediate and increasing expectations of financial analysts and shareholders to generate profits each quarter.

Objectives and Methods

This book covers the history and economics of the shift from industrial to institutional ownership in the United States and elsewhere and integrates the most relevant works from multiple fronts, especially economics, economic history, finance, industrial organization, forest management, and conservation. The primary focus is on the evolution of timberland ownership from government and/or private individuals to forest products firms first, and then to institutional investors, driven by the benefit-cost calculus of these owners, based on the returns and risks of timberland. In addition, it addresses the forest management implications associated with the rise of institutional timberland ownership. More specifically, the objectives of this book are as follows:

- to present how timberlands are valued and identify the drivers of timberland returns irrespective of ownership, and to demonstrate where sufficient market imperfection exists to enable one group of owners to generate higher returns and thereby buy timberlands from their existing owners;

- to explain the rise and fall of industrial timberland ownership and the shift from industrial to institutional timberland ownership in the United States;
- to present the historical roots of institutional timberland investment in the world and detail the contemporary development and financial performance of private equity in TIMOs and public equity in timberland REITs in the United States and elsewhere;
- to examine the management practices of institutional timberland owners and the implications for forest sustainability and forestry investing;
- to provide insights on the collaborative and competitive relationship among TIMOs and between TIMOs and REITs, as well as the relationship between institutional owners and a small number of remaining industrial timberland owners and nonindustrial private owners, which collectively are the largest group of timberland owners by area by a wide margin in the United States; and
- to look into the future of institutional timberland ownership in the United States and other parts of the world.

The research methods used in this book include a mixture of historical, institutional, analytical, comparative, economic, financial-economic, and forest-economic approaches. Forestry and forest resource management provide background information. The Faustmann-Hartman formula, one of the main forest-economic theories, is the base of timberland valuation and returns. The historical and institutional approaches used here offer a review of past and current events, allowing us to see a historical account of the changing timberland ownership and examine the effects of institutional factors such as tax laws and accounting rules. Collectively, these approaches help us investigate past events, important players, and the rules of the game related to timberland ownership.

The analytical and comparative approaches are based on benefit-cost analysis of various timberland ownerships. In particular, if an asset such as timberland is valued differently by different investors because of market imperfections, tax laws, and other institutional factors, a trade opportunity arises. These persons and organizations that facilitate such a trade have become the modern TIMOs and timberland REITs.

The financial-economic approaches include methods studying the risks and returns of timberland as an investment asset, arbitrage theory, and modern portfolio theory. The economic approaches include land rent theory, because those who can better manage land and create higher land rent will value timberland more and thus be able to buy timberland from others, as well as capital theory, because forests are a capital resource.

Plan of the Book

The next chapter presents the theoretical basis of this book—the market value of timberland and factors influencing timberland values and returns in a perfect market. It also presents examples of imperfection in timber and timberland markets that could be exploited by individuals, companies, and institutional owners and their managers.

Chapter 3 details the early history of institutional timberland investment and the pioneers and early practitioners of TIMOs in the United States who saw a business opportunity by bridging and facilitating the sales of industrial timberlands to institutional owners and becoming the managers of institutional timberlands.

Chapter 4 is devoted to the motivations of sellers and buyers of industrial timberlands in the United States. It reveals the pressure for forest products companies to sell their timberlands as well as the defensive moves made by some forest products companies in the early 1980s. It looks at the macroeconomic, market, valuation metric, regulatory, tax policy, and institutional (accounting rules) factors that contributed to the demise of industrial timberland ownership and the rise of institutional timberland investment in the 1980s and early 1990s.

Chapter 5 discusses the concurrent development of various TIMOs between 1988 and 2020. As earlier institutional investors secured excellent returns on their timberland investments and as timberlands were realized in the investment community as a promising alternative investment asset, demand for this asset class increased, which in turn called for more TIMOs. Direct institutional timberland ownership like that of the Harvard University Endowment is noted. This chapter also presents the forest management and sustainability implications of increasing institutional timberland ownership.

Chapter 6 documents the history and logic of creating timberland REITs in the United States, which began as a defensive move by forest products firms to avert possible hostile takeovers in the 1980s. This chapter also covers the dynamics and operation of timberland REITs and their advantages and disadvantages compared to TIMOs from the perspective of institutional investors. Finally, this chapter presents a comparison of the financial performance of TIMOs and REITs relative to other financial assets between 1987 and 2020, especially in the last ten to twenty years.

Chapter 7 covers institutional timberland ownership around the world. From the US perspective, global timberland investment has two dimensions: inward foreign investment (institutional investors from non-US countries investing in the United States) and outward foreign investment (US-based institutions investing in other countries). Globally, there is a third dimension, which is non-US institutional investment in timberlands on non-US soils, or non-US-based TIMOs investing in their own or other countries except the United States. This chapter also notes several publicly traded funds focusing on global timberland investment and the implications of increasing institutional timberland ownership for global forest sustainability.

The last chapter provides a summary of the causes and implications of institutional timberland ownership from economic, financial, and institutional perspectives. I conclude that high-quality timberlands make sense as a long-term investment and that, as long as the imperfection of timber and timberland markets exists, there are opportunities to earn abnormal returns by owning timberlands, and that institutions can continue to look for timberland investments in places where their return-and-risk calculus is more favorable. I also speculate where institutional timberland investment business goes from here, including the exchange of timberlands managed by TIMOs and timberland REITs, the emergence of secondary markets, more timberland investment tilted toward sustainability, forest-based climate solutions, and blended finance or private-public partnerships in timberland investment globally.

Appendix:
The Possibility of Generating Monopsonistic Rents through
Industrial Timberland Ownership

If one assumes that the capacity of a large forest products mill is fixed and
its demand for timber is Q, its timber supply comes from two sources: its
own timber (Q_i) and timber purchased from nonindustrial owners on the
open market (Q_n). The objective of the firm is to minimize the cost of timber,
$C(Q_i, Q_n)$, subject to $Q = Q_i + Q_n$. The price of its own timber is P_i, which may
be higher or lower than, or equal to, the price of timber on the open market,
and the price of timber on the open market is $P_n(Q_n)$.

$$\text{Min } C(Q_i, Q_n) = P_i\, Q_i + P_n(Q_n)\, Q_n \tag{1.1}$$

S.T. $Q = Q_i + Q_n$

Equation 1.1 can be changed to

$$\text{Min } C(Q_i, Q_n) = P_i\, (Q - Q_n) + P_n(Q_n)\, Q_n \tag{1.2}$$

Taking a derivative of $C(Q_i, Q_n)$ with respect to Q_n yields

$$\frac{dC}{dQ_n} = -P_i + P_n(Q_n) + \frac{dP_n}{dQ_n} Q_n$$

$$= -P_i + P_n(Q_n)(1 + \frac{1}{\varepsilon_n}) \tag{1.3}$$

where ε_n is the timber supply elasticity of nonindustrial timber on the open
market. Making equation 1.3 equal to zero yields

$$P_i = P_n(Q_n)(1 + \frac{1}{\varepsilon_n}) \tag{1.4}$$

If ε_n is equal to 0.5, then the marginal benefit of owning timber—or the
price of industrial timber—should be three times the price of nonindustrial
timber on the open market. In other words, the marginal revenue of industrial timber from industrial timberlands is three times that of nonindustrial
timber. This means that if forest products firms own some timberlands, they
can leverage and lower prices for nonindustrial timber on the open market
and generate monopsonistic rents. Empirically, stumpage (timber) supply
elasticities range between 0.3 and 0.6 in the US South and between 0.2 and
0.3 in the US Pacific Northwest (table 1.2).

Table 1.2. Stumpage supply elasticity estimates in the US South and Pacific Northwest

Source	Timber type	Elasticity estimate
US South		
Robinson (1974)	Pine stumpage	0.32
Adams and Haynes (1980)	Industrial stumpage (South-central)	0.47
	Nonindustrial stumpage (South-central)	0.39
	Industrial stumpage (Southeast)	0.47
	Non industrial stumpage (Southeast)	0.30
Adams et al. (1982)	Stumpage	0.41
Daniels and Hyde (1986)	Softwood and hardwood in North Carolina	0.27
Newman (1987)	Softwood solid wood	0.55
	Softwood pulpwood	0.23
Newman and Wear (1993)	Industrial sawtimber	0.27
	Nonindustrial sawtimber	0.22
	Industrial pulpwood	0.58
	Nonindustrial pulpwood	0.33
US Pacific Northwest		
Robinson (1974)	Douglas fir stumpage	0.11
Adams and Haynes (1980)	Industrial stumpage	0.26
	Nonindustrial stumpage	0.06
Adams et al. (1982)	Western stumpage	0.17
Adams (1983)	Private stumpage	0.15 to 0.34

CHAPTER 2
The Fundamental Value of Timberland

Timberland represents two economic factors of production: land, and capital that is embodied in timber. Land is essentially fixed in supply in a given region even though land use can change among agricultural, forestry, conservation, and urban uses. Timber as a capital asset can be built and depleted by harvests or natural events over time. Unlike agriculture, timber production is not labor intensive and can have a flexible harvesting schedule based on market conditions. Thus, forests are a factory as well as a product, so one can say that forestry is about appropriately managing timber stock and investment.

This chapter begins with the theory of timberland valuation at the smallest forest management unit—the so-called timber stand—first, bare timberland only and then timberland with trees, assuming that forestry is the highest-valued use for the land. I then list the sources of return for timberland investment and point out that market inefficiency may exist naturally or arise because of institutional and policy factors, which in turn influences timberland returns. Many ways that private investors, forest industry firms, and institutional investors and their advisers can exploit and benefit from market inefficiency are briefly noted in this chapter, while examples are presented throughout the book. Finally, because timberland values are drawn on a whole forest that typically has many timber stands, I provide some insights and comments on the practical application of timberland valuation at the forest level.

The Fundamental Value of Timberland[1]
The Value of Bare Timberland

As stated earlier, a given piece of land can change uses, which may significantly impact investment returns for specific properties. However, at the beginning of this analysis we do not consider land use change to higher-and-better uses (HBU) or nontimber benefits, or multiple uses such as mineral rights and hunting leases. Rather, we first focus on a representative piece of land on which timber production is the highest-valued use, both now and in the foreseeable future, and in a perfectly competitive market. For a representative unit of land that is being used for producing timber in perpetuity, the deterministic, expected value of a unit of bare land is

$$\text{LEV} = \frac{PQ(T, E)e^{-rT} - wE}{1 - e^{-rT}} \tag{2.1}$$

where LEV is land expectation value, P is expected timber price on the stump (stumpage price); T is rotation (harvest) age; Q(T, E) is expected timber volume or timber inventory at rotation age T and is a function of rotation age T and the level of initial silvicultural effort E; w is the unit cost of silviculture; and r is the continuous interest rate of a representative landowner.

Often, the term wE in equation 2.1 is thought to be the cost of establishing a timber stand, or the afforestation (reforestation) cost. However, all silvicultural costs that occur afterward can be discounted to the time when the timber stand is established using the landowner's interest rate. Subsequently, wE can be treated as the present value of all silvicultural costs that may occur during a whole rotation period between establishment and final harvest. Similarly, stumpage revenues received before a final timber harvest can also be added to the numerator of equation 2.1 to make the first term in the numerator the total present value of all stumpage revenues in the whole rotation. Here, the term "present" means at the time of timber stand establishment.

Equation 2.1 is the well-known Faustmann formula with an added silvicultural cost component.[2] If stumpage price and interest rate are constant, equation 2.1 is often used to solve for the optimal rotation age of the timber stand, T*, and the amount of optimal silvicultural effort, E*. Note that Q(T, E) is assumed to be invariable in various rotations, and no taxes are considered here for simplicity. In practice, stumpage price, interest rate, rotation age, and silvicultural efforts change over time and space and among landowners, and that is how market inefficiency can be exploited by investors and their skillful forest managers. Finally, our focus on a single unit of timberland is in keeping with tradition in the literature, but it requires the implicit assumption of constant returns to scale for timberland in a given land base. This returns-to-scale issue, like returns associated with land in different locations and of different quality, can also be exploited by skillful forest managers in practice.

We also need to consider nontimber benefits that the private landowner and the general public can enjoy from a forest that grows on this piece of land. Many of these nontimber benefits are captured by the public and are called environmental services. Yet the landowner may capture some non-timber benefits such as hunting and mineral rights leases, pine straw har-vesting, sales of conservation easements, and personal recreation. These nontimber benefits could accrue continuously (such as hunting leases), within an interval (such as pine straw collection), or as a one-time event (such as sales of conservation easements). If the nontimber benefits are a one-time event, for the private landowner it would mean an addition of non-timber value to the right-hand side of equation 2.1. If nontimber benefits are continuous and produced as a flow, then the nontimber values of a forest up to the rotation age T and silvicultural effort E are given by an accumulative function, N(T, E):

$$N(T, E) = \int_0^T n(t, E) \, e^{-rt} \, dt \tag{2.2}$$

where $n(t, E)$ is the nominal (nondiscounted) nontimber value at a given age t and given silvicultural effort E; and dt is the usual differential item in integration. If nontimber benefits accrue over an interval or only once, they can be treated as a special case in equation 2.2.

Adding the nontimber benefits would change equation 2.1 to

$$LEV = \frac{PQ(T, E)e^{-rT} + \int_0^T n(t, E)e^{-rt} dt - wE}{1 - e^{-rT}} \tag{2.3}$$

Equation 2.3 is an extension of the Hartman formula,[3] with the term repre-senting silvicultural investment added on. It represents the land expecta-tion value of a tract of forestland where a landowner captures both timber and nontimber benefits. Note that equation 2.1, which represents the value of bare land used for perpetual timber production only, is a special case of equation 2.3.

The Value of Timberland with Trees

If trees are already growing on the site, equation 2.3 needs to be modified further to include the value of trees. If the trees are ready to be harvested immediately and regeneration starts immediately after timber harvest-ing, then the net return from timber sales at the present must be added to the right-hand side of equation 2.3. If the trees are not yet ready to be harvested or have not reached merchantable age, or the landowner (or a new timberland buyer/investor) does not plan to harvest the timber now, the value of the standing timber and nontimber value must be added to the right-hand side as well. In addition, the right-hand side must be multiplied by e^{-rK}, where K is the number of years until the trees currently growing on the site are sold and harvested. The formula for determining the value of the asset—land with trees already growing on the site—is

$$AV = \frac{1}{e^{rK}}\left[PQ(K,E) + \int_0^K n(k,E)\,e^{rk}dk + \frac{PQ(T,E)e^{-rT} + \int_0^T n(t,E)e^{-rt}\,dt - wE}{1 - e^{-rT}}\right]$$

(2.4)

where AV stands for asset value that includes both bare land and trees, and $k = 1 \ldots K$. Note that the second term in the brackets is the accumulative future nontimber value at year K.

Equation 2.4 is the value of timberland with trees and can be decomposed into timber and nontimber value and (bare) land value:

$$AV = \frac{1}{e^{rK}}[PQ(K,E) + \int_0^K n(k,E)\,e^{rk}dk] + \frac{PQ(T,E)e^{-rT} + \int_0^T n(t,E)e^{-rt}\,dt - wE}{e^{rK}(1 - e^{-rT})}$$

(2.5)

In other words, the value of a timberland asset with trees consists of the present values of current timber and nontimber expressed in the first term of equation 2.5, and of bare land, shown in the second term of equation 2.5.

Note that equation 2.3, which presents the value of bare timberland with future timber and nontimber benefits, is a special case of equation 2.5. Furthermore, if K = 0, it means the standing timber is mature and immediately harvestable and that the right-hand side of equation 2.4 or 2.5 becomes simply the mature timber value, plus the bare land value (or LEV) that incorporates future timber and nontimber values.

Institutional timberland investors often buy and hold timberlands that vary from bare lands only, to lands with premerchantable timber, to lands with fully mature forests that can be harvested immediately. The total asset value of all timberlands for an institutional investor is thus the summation of all these lands with different timber species and age classes. In all events, the theoretical basis for the value of timberland properties and information related to the drivers of timberland returns contained in equation 2.5 remains valid.

Although I used an even-aged stand in the foregoing discussion, it should be noted that the Faustmann-Hartman formula is equally applicable in determining the optimal cutting cycle and growing stocks for an uneven-aged stand. In fact, if one treats as a silvicultural investment the merchantable volume left in an uneven-aged stand after timber harvesting in order to facilitate natural forest regeneration, it is apparent that the theoretical basis for maximizing the asset (land and forest) value in an uneven-aged stand is the same as that for maximizing the asset value of an even-aged stand. Thus, we have a unified theoretical approach to the management of even-aged and uneven-aged timber stands.[4]

Sources of Timberland Investment Returns

Equation 2.5 also shows how periodical (say, quarterly or yearly) total returns from timberland investment are generated in theory. The income returns during a period are represented in part of the first term on the right-hand side of equation 2.5 and are the sum of all cash income from timber harvests and nontimber product sales minus all concurrent silvicultural and management costs. The capital returns are shown in part of the first term (timber yet to be harvested) and the second term (bare land value).

Based on equation 2.5, we see that the total value of timberland and the returns from timberland investment are influenced by

(1) current and expected stumpage prices, P;

(2) timber volume, $Q(t, E)$, and the associated quality and dimension of timber;

(3) landowner's interest rate, r;

(4) optimal rotation age, T^*;

(5) total silvicultural cost, wE or E^* if one assumes that the unit silvicultural investment w is market driven and not influenced by individual landowners;

(6) nontimber benefits ; and

(7) age of the existing trees, which is $T^* - K$. When the current age is greater than T^*, the stand is often treated as a mature timber stand that is ready to be harvested, just like a stand with age T^*.

For completeness, I would add three other factors that are not included in equation 2.5: (8) HBU, (9) taxes, and (10) accounting rules. An HBU implies that timberland is going to be converted to a different use and should be valued accordingly. Taxes and accounting rules are institutional arrangements. And, as we will see, differential tax treatments for the same timberland among different owners and accounting rules are a major cause of the shift of industrial timberlands to institutional ownership in the United States.

Figure 2.1. Softwood sawtimber and pulpwood stumpage prices in the US South in 2020 constant dollars, 1955–2020

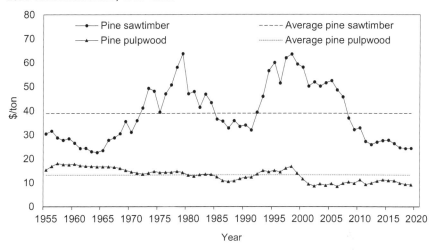

Sources: Timber-Mart South Inc., Journal of Southern Timber Prices (Athens, GA, various years); Ulrich (1990).

Now I discuss the factors that influence timberland value and returns, identify the sources of potential market inefficiencies, and point out possible avenues for smart landowners and investors to profit from these inefficiencies.

Stumpage Prices

In the deterministic Faustmann formula, stumpage prices are assumed to be constant in the long run. As products and factor markets develop and timber supply and demand shift in response to macroeconomic and microeconomic conditions, stumpage prices change accordingly. Unless there is Soviet-style social and economic planning, it is impossible to generate a constant stumpage price over time. Nonetheless, one can think of such a constant and long-run price as an average stumpage price over many decades. Because stumpage prices vary significantly over time and space in the short run, smart investors can benefit by identifying and choosing to invest in timberlands with appropriate species, age class, composition, and growth potentials at the right time and location, and by applying suitable management and harvesting options.

Figure 2.1 shows the real softwood sawtimber and pulpwood stumpage prices in the US South from 1955 to 2020. While real softwood pulpwood prices were relatively stable, real softwood sawtimber prices fluctuated greatly in the last sixty-six years. Unlike agricultural products, which are often subject to supply-side effects, forest products and stumpage markets fluctuate mainly because of shifts in demand. Thus, we see a rise in sawtimber stumpage prices from 1965 to around 1980, until the demand for forest products (and thus stumpage) was curtailed by a severe recession. And the curtailment of Canadian softwood lumber imports, which increased US demand for domestically produced lumber after December 30, 1986, again sent US softwood lumber and sawtimber prices on an upward trajectory.[5]

Supply limitation or expansion also influences stumpage prices. The listing of the northern spotted owl as a threatened species reduced production from public lands and raised stumpage prices from private lands. On the other hand, the Conservation Reserve Program and other government programs, which subsidized private landowners to plant trees on 25 percent more acreage between 1986 and 1995, have contributed to the prolonged slump in sawtimber stumpage prices in the US South since 2007.[6]

Stumpage prices also vary in different regions with identical species or similar species for the same uses. For example, Douglas fir and southern yellow pine sawtimber are both used to produce softwood lumber. Partly because the quality of lumber produced from Douglas fir (straight and easy to nail) is better than that from southern yellow pine, among other things, Douglas fir stumpage prices are typically higher than southern pine stumpage prices. Yet the reverse was true in the mid-1980s (fig. 2.2). Observing

Figure 2.2. US Pacific Northwest and southern sawtimber stumpage prices in 1982 constant dollars, 1960–1985

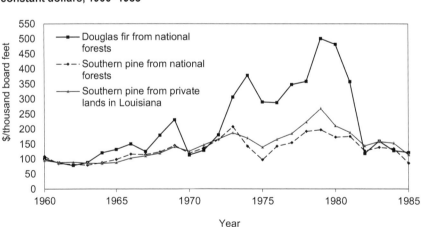

Source: Ulrich (1990).

this discrepancy and believing that the deep discount in timber values in the Pacific Northwest would not last long, some institutional investors invested mainly in the US Pacific Northwest in the late 1980s and received superb returns a decade later.

Since the early 1990s, forest certification has been promoted as a means of achieving forest sustainability. In particular, certain segments of the forest products market around the world have demanded third-party certification for forest products through chains of custody all the way to forests. It was said that certified woods would lead to a price premium as well as market access. To the extent that timber from a certified forest can generate a price premium that more than covers the cost of certification, forest owners should pursue certification of their forests. Product differentiation is thus another market inefficiency that forest owners and managers can exploit.

Similarly, thanks to technological breakthroughs, some species previously considered "weeds" have started to be used for making paper and engineered wood products. Also, large amounts of renewable green energy in Europe and North America are being produced based on residual wood biomass, increasing the competitiveness of wood-based energy compared to fossil fuels. As a consequence, there has been a growing market and a newly established market price for residual wood and weed species that were previously valueless. Further, carbon-related regulation and the desire for a bioeconomy based on material circularity are in place (France already has such regulations for public buildings) or are planned for house building, which is likely to result in increasing demand for cross-laminated timber (CLT) and similar solutions, driving up future timber prices.[7]

A final note on stumpage prices is that timber markets are heterogeneous and extremely varied, unlike commodities sold in supermarkets whose prices vary little in some jurisdictions. For example, in any typical bid for timber on a good tract of timberland in the US South, the highest bid often exceeds the lowest by more than 100 percent. Sales methods and the illusion of knowledge (lack of information, a form of market imperfection) on the part of buyers are two of the many reasons that timber markets behave this way.[8] Similarly, stumpage prices equal delivered log prices minus logging and transportation costs. In some markets, the availability of logging contractors impacts stumpage prices as well as the ability of landowners to sell their timber.

Figure 2.3. Growth in volume and stumpage value of a forest as it ages

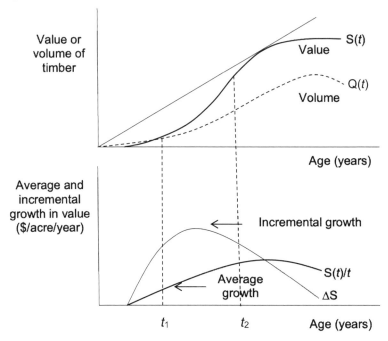

Adapted from Zhang and Pearse (2011). Note that the incremental growth in stumpage value is much higher than the average growth in stumpage between t_1 and t_2.

Timber Volume

The volume of a forest typically grows as a dashed sigmoid curve, as shown in the upper quadrant of figure 2.3. Its slope increases up to an inflection point and then decreases, a growth pattern frequently seen in biology. The volume continues to increase as long as the annual increment of growth, which diminishes after the inflection point, exceeds the increasing losses due to insects, diseases, and natural mortality as trees approach their biological age limitation.

Along with natural biological growth, trees become bigger, and more valuable products can be manufactured from larger timber. For example, large-dimension lumber and high-quality veneer can be milled only from large logs, and larger logs have a larger proportion of clear grain. This growth of trees to prescribed dimensions or sizes, which enables higher-valued products to be made from them, is often called ingrowth. In economic terms, it can be called appreciation of trees.

Table 2.1. Nominal stumpage prices by product in current dollars in the southern US, 1980, 2000, and 2020

Product	1980	2000	2020
Softwood pulpwood ($/ton)	4.35	7.78	8.44
Softwood chip-and-saw ($/ton)	10.45	27.54	16.11
Softwood sawtimber ($/ton)	16.53	39.10	23.33
Softwood ply logs ($/ton)	20.29	40.10	28.32

Source: Timber-Mart South Inc., Journal of Southern Timber Prices (Athens, GA, various years).

For example, in the US South, trees first reach pulpwood size, then chip-and-saw (small-diameter sawlogs that can be used as either pulpwood or sawtimber depending on the market prices of paper and lumber), sawtimber, and ply logs (which are used for making veneer and plywood), each having a progressively higher price per unit of wood. Table 2.1 presents the current stumpage prices for softwood pulpwood, chip-and-saw, sawtimber, and ply logs in the southern United States in 1980, 2000, and 2020, respectively. It shows that in good or bad markets, the prices of an upper-level product are always higher than those of a low-level product for the same amount of wood.

This price jump associated with ingrowth, plus the fact that large timber can usually be harvested at a low cost per unit of wood, means that stumpage revenues typically increase with age in a pattern that is similar to, but often steeper than, the volume growth curve, as shown in the curve S(t) in the upper quadrant of figure 2.3. Put another way, a forest's growth in value (stumpage revenues) normally exceeds its volume growth because trees grow not only in cubic volume but also into more valuable product categories.

Nonforesters may initially have difficulty comprehending what this dynamic of tree growth and ingrowth could mean for timberland investment. In a legal case in southern Alabama that I was involved in, a business valuation expert testifying in court for a timberland company used a 5.3 percent growth rate that was provided to him by the management of the company. To demonstrate that this expert did not know what this 5.3 percent really meant and that it was inappropriate for him to simply take this number from the management of the company and use it, the opposing attorney asked the valuation expert this question:

"Let's do this. Let's take that 5.3 percent. That's how fast you say trees are growing. That would be the annual biological growth rate?"

"That's management's calculation," the expert replied.

"All right. And tell me, if the tree was one-foot tall when you planted it, how tall would it be in thirty years?" the lawyer asked.

The answer would be 4.7 feet, a ridiculously short tree in the state of Alabama. Yet this expert did not know how the 5.3 percent was calculated or how it should have been properly used.

A lack of understanding of the growth pattern in timber volumes and values was rather common among nonindustrial forest owners years ago and is still present to some degree today. More importantly, many of them (and even some industrial timberland owners) do not actually measure the volume (and thus do not have a value) of timber on their land when they try to sell their timber or timberland. This situation of uninformed sellers in timber and timberland markets existed in the United States at least until the 1990s, allowing some timberland companies and brokers to profit from this inefficiency. Georgia Timberlands Inc. was such a company that was able to take advantage of this situation for many years and is used as an illustrative case study here.[9]

* * *

Georgia Timberlands was founded by George Peake Sr. in Macon, Georgia, in 1947. Before that, Peake was in the pecan-producing business in Eufaula, Alabama. When the pulp and paper industry moved to the South, the Peake family seized the opportunity to diversify into the timber brokerage and land management business. The family moved its business to Macon, Georgia, the site of the first inland paper mill in the South. Gradually Georgia Timberlands accumulated timberlands, and by the mid-1970s, it owned some 202,000 acres in Alabama, Georgia, Florida, and North Carolina. Between 1950 and 1980, it also probably "flipped" more than a million acres of timberland from farmers and other nonindustrial timberland owners, mostly to forest products companies.

How did Georgia Timberlands flip and accumulate timberlands? Whenever it heard about or saw a piece of timberland for sale that was of a certain size (200–400 acres being ideal), it would send a forester to go and inspect the timberland, eyeball the timber inventory, and assess its accessibility and timber growth potential. If timber accounted for at least 60–70 percent of the timberland value, Georgia Timberlands would make an offer within days at 85 percent of its estimated market (or intrinsic) value, with the option of increasing the price a couple of percentage points if it did not close the deal within three months. Often, the sellers would agree to sell, leaving Georgia Timberlands at least a 15 percent profit margin. In nearly all cases, the sellers did not know the timber inventory, as most of them did not want to spend the money to hire a forester to do a timber cruise when they decided to sell their lands, nor did they have an estimated value of their timber and timberlands.

In the meantime, Georgia Timberlands would hire a trusted forestry consultant to do a detailed timber cruise to assess the actual timber inventory and decide whether it wanted to flip the timberland or keep it for itself. If Georgia Timberlands decided to keep it, it would sell the mature timber after purchase. Thus, the accuracy of the forestry consultant's timber cruise would be known shortly afterward and credibility would be solidified. If Georgia Timberlands did not want to keep the land, it would also immediately start to seek a buyer, flip the property, and get a profit margin much higher than the normal brokerage fee of 4–6 percent. Typically, by making five to ten phone calls in a week, Georgia Timberlands would get a buyer.

Georgia Timberlands could make a quick offer because it had a $5 million line of credit from a local bank, and if needed, it could get a long-term loan from Travelers Insurance Company. Furthermore, Georgia Timberlands knew the timberland markets in its operating areas well and had a list of potential buyers, which were mostly forest products firms that were accumulating timberlands before the early 1980s.

* * *

Georgia Timberlands' experience shows a type of information inefficiency in timber and timberland markets: uninformed sellers and lack of market information. Some sellers do not do timber cruises because they want to save money, and they do not know timberland markets well. When they are offered a large lump sum of money, they often take it without knowing the amount of timber they have or the market value of their timberlands.

When some nonindustrial forest owners hire professional foresters to measure what they want to sell in either just timber or timberland with trees, they often ask foresters to use a sampling method (5–10 percent) instead of 100 percent measurement. Because the sample might not be representative of the whole and can be measured incorrectly (woods work is hard), an underestimate is normal rather than an exception. James Vardaman noted that in a timber sale, a timber buyer made a 10 percent sample inventory and estimated the volume to be 393,717 board feet. His forester measured

100 percent of the trees and found 502,200 board feet. Shortly thereafter, the timber was cut and hauled to a mill, where the volume was measured at 510,143 board feet. If the mill measurement determined the correct volume, the errors of these estimates were 22.8 percent for the 10 percent sample, and 1.6 percent for the 100 percent sample.[10]

Interest Rates

US capital markets are relatively efficient. Normally an interest rate used in an investment project has three components: a safe or risk-free rate represented mostly by a short-term US government bond rate, a broad market rate, and an adjustment to the broad market rate because of idiosyncratic project factors. In terms of a forest investment, the latter factors include location, topography, market conditions, species composition, and age classes, among others.

Several variables are candidates for proxy for the broad market rate or returns for alternative investments for capital. Returns from long-term corporate bonds and government bonds are the ones most cited, but not returns from common stocks. As we will see in chapter 6, empirical studies show that timberland returns are not related to broad stock market returns. On the other hand, long-term corporate bond returns capture timberland returns in the form of biological growth regardless of the markets. Furthermore, I have heard from various institutional investors and TIMO managers that their real (inflation-adjusted) discount rates are "real 10-year U.S. government treasury yield plus a couple percentage points."[11]

The average rate for all or most new timberland investment in the United States at a given time indicates the prevailing market rate for US timberland investment. Normally, the capital markets for timberland investment are a small subset of, and thus are likely to be less efficient than, the broad US capital markets. Anecdotal evidence also shows that this is the case in practice, so that some investors can take advantage at the expense of the others.

The best case in point is that industrial timberland owners often have an overall cost of capital at the corporate level that is different from their cost of capital for timberlands. Actually, this is one of the main reasons that industrial timberland owners started to divest their timberlands, a focus of this book. We will discuss this point further in chapter 5.

Another example is that the single average rate of interest used for timberland investment often does not reflect the maturity of the standing timber. As shown in figure 2.4, typically the average (or required) rate of return

Figure 2.4. US Treasury yield and timberland return expectations, 2010 and 2018

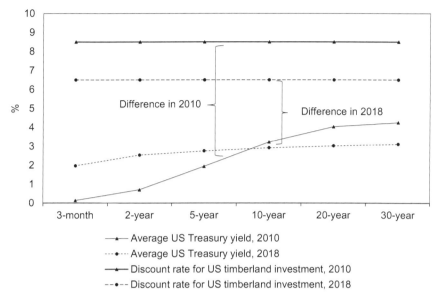

- Average US Treasury yield, 2010
- Average US Treasury yield, 2018
- Discount rate for US timberland investment, 2010
- Discount rate for US timberland investment, 2018

Sources: Annual average US Treasury yield for various maturity periods is from Federal Reserve Bank of St. Louis, accessed June 20, 2020, https://fred.stlouisfed.org. The discount rate for timberland investments is from conversations with TIMO managers. This graph shows that the average discount rate for timberland investments is a few percentage points higher than the ten-year US Treasury yield. However, the "few percentage points" vary over time. More importantly, if the discount rate for timberland investments is by and large fixed at a given time irrespective of forest age classes and maturity, there will be an inefficiency in timberland investment markets. Forests that are nearing maturity and that can generate revenue soon should have a lower discount rate than forests that can generate revenues only many years later, similar to how US Treasury bonds with a longer maturity have a higher yield.

for timberland investment is fixed at a given point in time, irrespective of the maturity of standing timber. Yet long-term government bond rates are typically higher than short-term government rates. So theoretically, timberlands with younger forest stands should have a higher required rate of return than those with more mature forest stands that are expected to yield income soon. If the US government yield curve is relatively flat, as in 2018, this may not matter much. If the yield curve is steep, as in 2010, the inefficiency in capital markets for investment in timberlands with different forest maturity is exacerbated.[12]

All else being equal, the investor with the lowest interest rate will win the bid in a timberland sale. For example, conservation-minded investors may have a lower required rate of return because their main motivation is conservation. Indeed, some funds focusing on forest conservation are able to acquire certain industrial timberlands and put them under conservation easements. Similarly, tax-paying institutions have a higher required rate of return than tax-exempt institutions. The existence of differential interest rates and capital market inefficiency in timberland investment allows some investors to profit from timberland ownership.

Optimal Rotation Age and Silvicultural Investment

Both optimal rotation age and silvicultural investment are management issues that have intrigued forest managers, forest economists, and policy analysts for over a century as a classical problem of investment analysis. Most trained foresters and forest managers should have the skills and expertise to make sound decisions on these matters. In practice, they are such complicated matters that even large landowners and their forest managers make various inefficient decisions.

The difficulty in making the most efficient decisions on rotation age and silvicultural investment arises because they are related to all the factors on the right-hand side of equation 2.3. Thus, an optimal rotation age, for example, is influenced by future stumpage prices, expected timber volume, the landowner's interest rate, silvicultural investment, and expected nontimber benefits. Similarly, optimal silvicultural investment is influenced by stumpage prices, interest rate, unit silvicultural cost, nontimber benefits, and rotation age. The future values of most of these variables can be assessed or estimated based on current and historical data or using sophisticated models. In many cases, these assessments turn out to be inaccurate. For example, very few forecasted that stumpage prices in the US South would stay low in the last decade or so (fig. 2.1). When these variables vary (and they tend to vary a lot) from their historical means, decisions on optimal rotation age and silvicultural investment need to be adjusted accordingly.

Complicating the matter is that timber production takes many years. A typical nonindustrial forest owner who owns less than one hundred acres of forestland in the United States may harvest timber only once every ten to twenty years. A fund for institutional timberland investment may last only ten to fifteen years, and hence fund managers must decide whether to harvest existing timber more or less intensively, and whether to invest in regeneration after harvesting. The latter will surely depend on whether managers believe they can recoup their investments if they decide to sell the timberlands in the fund in a few years. A private sawmill owner who also owns timberlands in Alabama and Georgia does not reforest after timber harvesting. Instead, the timber proceeds are used to subsidize sawmills and purchase more timberlands. Based on my calculations, this landowner's decision to not reforest after timber harvest could reduce the land expectation value by 30–40 percent. In other words, this no-reforestation decision can be justified only if the landowner can purchase new lands at a 30–40 percent discount.[13]

I must point out that a decision on optimal rotation age is also related to timber volume growth, as noted above, as well as the theory and practice of forgoing current income and storing value on the stump if log markets are weak. Under some circumstances, and perhaps most, storing value makes a great deal of sense, especially for small and nonindustrial private forest owners that harvest timber only once every ten to twenty years. For large, industrial or institutional timberland owners, however, the calculus is a bit more complex, even though the logic of delaying timber harvesting is the same for all landowners—the forgone income today is expected to be compensated by an increase in value from a combination of biological growth, ingrowth, and real price appreciation. But what if real timber prices do not recover, or what if they stay low for a longer period than expected? If so, it is possible that the increase in future income or value may not exceed the forgone income by at least the landowner's cost of capital.

Furthermore, large landowners may have other practical limitations in an inefficient market. They may not be able to ramp up their harvests when timber markets are strong because of a shortage of logging contractors. Conversely, they may have to harvest some timber to keep these logging contractors in business when timber markets are weak. Forest products mill capacity is another factor. Even in good timber markets, mills may take only so much timber at any given time, forcing large landowners to curtail their timber harvests. On the other hand, when a major paper mill in Courtland, Alabama, was shut down in 2014, timber prices plummeted in the region, and some of the value stored on the stump was washed away and could not be recovered. This is why some institutional timberland owners have resorted to owning a manufacturing facility that processes their timber.

Nontimber Benefits

As stated, most landowners can capture only a portion of nontimber benefits, and the rest are public goods that include mostly environmental services such as biodiversity, carbon sequestration, soil and water conservation, and landscape scenery. Here we focus only on the portion that landowners can capture now and in the future from possible securitization of some public goods. Most nontimber benefits are generated from the whole forest property rather than a representative tract of it, and thus nontimber benefits may best be considered as multiple uses on the whole property.

In the United States, private timberland owners can generate income from hunting leases, pine straw harvesting, fuelwood collection, resin and flower gathering, mineral rights, and sales of conservation easements and recreational rights to the public. For example, Forest Capital Partners was paid $140 to $200 per acre by the state of Minnesota for an easement that has since allowed public access to the 140,000 acres of forestlands it managed in the state. This payment alone was about 20–30 percent of the timberland value.[14]

The development of markets for some public goods is promising. For example, the United States has wetland mitigation banking (including streamside) and conservation banking programs in which landowners can sell wetland mitigation and endangered species conservation credits and thus generate income. The state of California has developed a "cap-and-trade" greenhouse gas emission reduction strategy, which has a compliance offset protocol for US-based forest projects (USFP). This USFP has laid the groundwork for both producers and buyers to begin transacting in compliance-driven carbon reduction through forest projects such as afforestation, reforestation, improved forest management, and avoided conversion.[15] Similarly, other companies in the United States, such as NCX, are developing a forest-based carbon offset market called Natural Capital Exchange (NCX).[16] The development of these "payment-for-environmental-services" markets will help forest landowners generate additional income from their forests if they can overcome associated transaction costs. One could also argue that forest certification is a securitization of environmental services.

The Ages (and Species Composition) of Existing Trees

Forests with different ages have different growth potentials in volume and value. Because growth in forest value is not strictly linear or exponential over time, the ages of existing trees on a timberland are a significant factor in determining the rate of returns of terminal timberland investments with a duration of ten to fifteen years. James Vardaman stated, referring to the increase in value in southern forests:

> Within certain limits, a young tree is a highly profitable investment, and an old one should be liquidated. For instance, a 6-inch pulpwood

Figure 2.5. Estimated value of a premerchantable loblolly pine timber stand

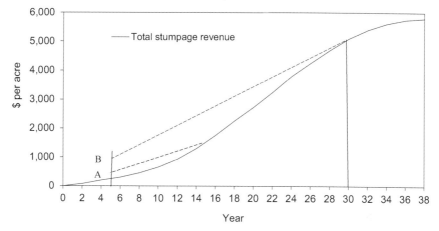

Adapted from Zhang and Pearse (2011). This graph shows the difference between the discounted value of a five-year-old timber stand at age thirty, when the stand captures the full spur in stumpage revenue growth and is ready for a final harvest (valued at B), and that at age fifteen, when the stand just becomes merchantable (valued at A), using an identical discount rate.

tree, by growing 4 inches in diameter, often increases in value 433 percent, and this growth may take place in as few as 12 years. On the other hand, a 16-inch sawtimber tree, by growing the same 4 inches, often increases in value only 71 percent, and this growth can take place in 12 years only under good management.[17]

This increased growth in value is associated with the aforementioned biological growth and appreciation of trees as a result of ingrowth. Thus, many TIMOs have tried to buy timberlands with trees that will have increased growth in value, which is the period between t_1 and t_2 in figure 2.3.

A related factor is that growth in value often trickles down to the valuation of premerchantable timber stands. Figure 2.5 shows that the estimated value of a five-year-old loblolly pine forest approximately doubles with one identical interest rate if one uses the projected income based on most of its products being sawtimber at age thirty instead of the income from pulpwood at age fifteen. Thus, the age of existing trees is an important factor in determining the value and potential returns of timberlands.

So far, we have assumed that our given tract of forest has no variation in species composition. However, different tree species have different growth rates, uses, and markets, allowing investors to select and specialize—just as investors choose to invest in forests of different ages. Indeed, some private individuals and institutional investors have focused on investments in timberlands with forests that have desirable ages as well as species composition and management style (for example, even aged vs. uneven aged).

Higher and Better Uses (Land Use Conversion)

An HBU for timberland is different from multiple use and implies a change in land use from timberland to another use. Multiple use, on the other hand, considers and incorporates all nontimber benefits of timberland.

Some agricultural and forested lands in the US South are interchangeable depending on the land rent that each use can bring to their owners. Some timberlands, because their conservation value is high, can be bought out by government or conservation organizations. Most often, higher and better use for timberland means conversion to urban, residential, commercial, or recreational use, putting it out of timber production. Residential use includes homes associated with urban expansion and summer houses for wealthy families. Commercial use includes power lines, electronic and communication towers, solar energy production (solar panels), storage facilities, and others. This dynamic change in land use goes on irrespective of who owns the timberland. It just so happens that some timberland owners are more proactive in seeking HBU opportunities, even for just a small portion of their whole ownership portfolio. Industrial owners, on the other hand, are known to focus more on the timber supply from their lands.

Taxes

Paying taxes reduces the net (after-tax) returns for landowners. Timberland owners in the United States pay mainly property taxes and income taxes. While property taxes may vary in different states and counties and thus are embedded or priced in timberland values in various locations, income taxes vary according to different ownership and ownership structures—corporate versus family ownership, different corporate structures such as C corporations and S corporations, and level of income.

Particularly, and as noted earlier, all C corporations, which most forest products companies are, pay income tax twice, once at the corporate level when timber incomes are generated and once at the shareholder level when shareholders receive dividends. On the other hand, general partnerships, limited liability partnerships, and REITs are "a corporate vehicle without corporate taxes" and are taxed only at the shareholder level. All these tax rules have created an uneven playing field for different landowners, which is one of the institutional market inefficiencies that causes landownership shifts in the United States: all else being equal, those who pay less taxes will buy timberlands from others who pay more taxes.

Accounting Rules

The values of timber and timberlands are kept in financial statements of public companies in different ways based on the prevailing accounting rules in different countries and jurisdictions. In Oceania and Europe, fair market value is applied to timber and timberlands as well as other assets. So, all potential buyers see the same fair market value for timberlands on the financial statement of the seller. Of course, the buyer and seller can value the timberland differently. Otherwise there would not be any transactions.

In the United States, generally accepted accounting principles (GAAP) prevail. Under GAAP, publicly traded forest products companies carry timberland or other assets on their books—financial statements—at cost or market value, whichever is lower. So, if a forest products company bought a piece of bare land and planted trees on it in 1980, it could register the asset at only the cost of land and planting expenses. That might have been on the order of $200 to $300 per acre in the US South at that time. When that timber matured in 2010, it could have had a fair market value of more like $2,000 to $2,500 per acre, while the land value might have risen to perhaps $700 per acre. However, the company could not revalue the land or mature timber in compliance with GAAP between 1980 and 2010. As a result, the book value of the timberland would have been only a fraction of its fair market value in 2010. Similarly, if a timberland asset with mature timber was purchased long ago, and mature timber was "depleted" through timber harvesting and new trees grew back, the book value would also be a fraction of the fair market value because of inflation, tree growth, and real price appreciation in timber.

This undervaluation of industrial timberland under GAAP made some people regard timberlands as "deadweight" on corporate balance sheets in the 1980s. More importantly, it gave corporate financiers an opportunity to buy a forest products company and sell the timberland and other assets at fair market value. Furthermore, when forest products companies realized that they were under the threat of possible takeover, they could start to aggressively sell some timberland assets. Finally, even if forest products companies were not threatened, they could sell some of their timberlands when they saw that they could not meet their financial goals at the end of their financial years. Indeed, selling undervalued timberland was a quick way for forest products companies to reach their year-end financial goals, allowing company executives to retain their bonuses and jobs. Many forest products companies did so in the 1980s and 1990s.

In short, institutional arrangements regarding taxes and accounting rules also contribute to the values and returns of timberland, in addition to market, biological, and managerial factors.

Timberland Valuation at the Forest Level

Although the timberland valuation and investment model presented above is theoretically sound and represents a major contribution of forest economists such as Martin Faustmann to general economic theory long before mainstream economists such as Irving Fisher and Paul Samuelson, it is a stand-level model. Often investors—individuals, industrial firms, or large institutions—make investment decisions at the forest level.

Simply stated, a timber stand is a contiguous group of trees sufficiently uniform in age-class distribution, composition, and structure, and growing on a site of sufficiently uniform quality, as to be a distinguishable unit.[18] As the smallest forest management unit, a timber stand may be called a compartment or subcompartment in some countries. Rarely, a forest has only one stand, but usually, a forest that is defined by ownership or a natural boundary often consists of multiple timber stands with varying ages, species composition, and site productivity. Furthermore, an investment opportunity may include multiple forests located at different sites with different natural and market conditions.

Therefore, investment opportunities are analyzed and valued at the forest level. One way to do a valuation at the forest level is to use the simple mathematical promise that the sum of the parts (stands) approximately equals the whole (forest). Yet, as Aristotle said, "The whole is greater than the sum of its parts." So, while the stand-level method is valid and may be acceptable because each stand must be managed according to the particular inventory, capability, and circumstances that drive its fundamental value as outlined above, I must point out that the estimation results based on a forest-level valuation approach can differ substantially from those of a stand-level valuation approach for two reasons. First, major decisions about access development such as road building, silvicultural investment and protection, harvest rate, nontimber benefits, and measurement of economic performance normally apply to the whole forest. In other words, individual stands must be managed in ways that contribute to the objective of the whole forest even if these management decisions may be suboptimal as far as individual stands are concerned.

Second, there are complementary and competing relationships in the management of many stands. That is, there are economies of scale and dis-economies of scale when dealing with a whole forest. For example, applying silvicultural treatments in a large forest could incur a lower unit cost for multiple stands than would treating individual stands separately. Similarly, forestry activities such as fire protection and pest control need to be coordinated over many stands. A large forest with thousands or millions of acres could be offered at a price with some discount, at least at the beginning of large industrial timberland sales. This is good for potential buyers. On the other hand, disposing of a large forest of millions of acres could take many years, implying a diseconomy of scale in large timberland sales.

In practice, each investment opportunity on a large forest asset is evaluated and underwritten with a modification of the sum-of-the-parts approach by investors. These investors (and/or their consultants or managers) often use a forest inventory, a set of growth and yield models, and a price forecast as well as their cost of capital to develop a forest-level harvest schedule. Numerous optimization models are used for this purpose, but the Woodstock software developed in Canada is standard. This kind of software respects such constraints as forest practice regulations, sustainability, and market absorption. The harvest schedule is then fed into a cash-flow model (including taxes) to calculate the net present value of the forest asset, or, more usually, the internal rate of return at an assumed purchase price, along with an assumed sale price for the asset in the future. Often, HBU, option value, and uncertainty are also considered in this process. This whole valuation process requires adequate knowledge and skill in forestry, economics, finance, and programming as well as sound professional judgment and is critical in timberland investment.

CHAPTER 3

Emerging Institutional Timberland Investment and the Rise of TIMOs in the United States, 1952–1987

This chapter covers the historic roots of institutional timberland investment and the emergence of TIMOs in the United States in the modern era—the first half of the last sixty-plus years. I first distinguish between private individual and institutional timberland investment, describe the evolution from the former to the latter, and present the multiple historical origins, pioneers and early practitioners, and their investment propositions for the TIMO business in the United States. The conclusions of this chapter are that institutional timberland investment has evolved from private timberland investment and that policy developments and institutional factors have aided the rapid rise of institutional timberland ownership. A few private and institutional investors who received good financial returns from timberland investment started the prototype TIMO business in the 1960s and 1970s. They were joined by a few more entrepreneurs who saw a business opportunity in helping institutions obtain timberlands and advising them, thereby creating a demand for timberlands in the 1980s.

Private Individual versus Institutional Timberland Investment

Private individual timberland ownership has likely existed since private ownership of assets was allowed. The earliest example of institutional timberland ownership is the extensive timberland ownership by churches in Europe.[1] However, the distinction between private individual and institutional timberland ownership in the modern era is not very clear for three reasons.

First, as noted earlier, institutional timberland ownership as it is defined today has evolved from private individual timberland ownership, as institutions became interested in timberland investing after learning from the experience of some private individual forest owners. Second, some individual private timberlands were put into trusts managed by the trust departments of banks, which were institutions themselves. Commercial and investment banks and private forestry investors also began raising money from high-net-worth individuals and institutions such as pension funds to invest in timberlands from the 1960s to 1980s. In either case—whether private owners willingly put their timberlands under the management of an institution, or an institution succeeded in raising money from private individuals to buy timberlands—there was an institution that advised and managed the timberlands. Finally, some wealthy individuals who invested in timberlands had their own investment management teams and advisers, which were often referred to as family offices and acted somewhat like institutions.

Thus, there exists a continuum between private individual and institutional timberland ownership, and the line between them is blurred nowadays. Here institutions include both tax-exempt institutions such as pension funds and endowments and tax-paying institutions such as private funds and insurance companies' own timberland investments.

It is important to distinguish between the mortgage lending business and the equity position in timberlands held by banks and insurance companies. The mortgage lending business, in which banks and insurance companies serve as creditors to facilitate the purchase of timberlands by others, is not a form of active and direct ownership, because the creditors are not involved in timberland management and do not directly control the timberland assets. Even if the creditors sometimes end up with timberlands in case of default or bankruptcy by the borrowers, this is a passive ownership of timberlands that tends to be short-lived, because the creditors often sell the properties to cover their loans and do not intend to directly hold, own, and manage them for long. On the other hand, the equity position held by banks and insurance companies is a form of direct ownership because these institutions control and manage the timberlands and take the associated returns and risk. But, as we will see later in this chapter, the experience of mortgage lending provided an opportunity for banks and insurance companies to better understand the financial characteristics of timberland investment, encouraging them to subsequently take an equity position in timberlands.

I should also point out that, while all timberland REITs are organized and operate in a similar fashion, TIMOs are quite diverse in their organizational structure, function, services, and operation. While all TIMOs advise their clients on timberland investments, each TIMO may do one or more of the following: promote and market timberland investments, manage timberland, engage in acquisitions and dispositions, and act in a fiduciary capacity. Some of them do all these things as integrated firms. These TIMOs may have individual institution-specific separate accounts as well as pooled (commingled) accounts with funds that have capital from multiple institutional investors. For institution-specific separate accounts, the duration of the investment is dictated by the institutional investor and can be open ended, whereas pooled accounts generally have a fixed term, often ten or more years with the possibility of extension.

Other TIMOs such as Wagner Forest Management Ltd. do not actively market and raise monies but focus on acquiring, managing, and selling timberlands for both private individuals and institutional investors at their request. Finally, some "quasi TIMOs," much like real estate brokers and/or forest investment advisers, facilitate direct investment in timberlands by private or institutional investors and help these investors hire on-the-ground forestry consultants to manage their timberlands. R&A Investment Forestry, which ceased operation after the retirement of its principal, was a good example of a quasi TIMO.

Timberland ownership through TIMOs is mostly private equity structured in the form of funds that often are organized as general partnerships, limited liability partnerships, or private REITs. Most large timberland REITs are publicly traded and their returns are more closely correlated to the returns of general stock markets than are the returns of private equity timberlands.

When forest products companies owned some seventy-one million acres of timberland in the United States in the 1970s and 1980s, owning the stocks of these companies included some degree of timberland investment. However, timberlands typically represented only a small portion of the forest products firms' total assets. Furthermore, as noted earlier, under GAAP, the values of timberland assets were understated, which is one of the main causes of declining industrial timberlands and rising institutional timberland ownership through TIMOs and timberland REITs. Third, as noted in chapter 1, there can be conflicts when forest products plant managers want to restrict the transfer prices of timber harvested from their own lands. Thus, institutional timberland ownership through TIMOs, timberland REITs, or direct investment is much different from institutional ownership of integrated forest products companies' stocks. The former is focused purely on timberland and recognizes its fair market value, while the latter is influenced by a multitude of other factors, including accounting conventions.

Emerging Institutional Timberland Investment and the Rise of TIMOs in the United States

Institutional timberland investment in the United States began with small-scale, exploratory direct investment by a few institutions, first via debt and then through both debt and equity, and by some wealthy individuals, mostly via equity, between the 1950s and 1970s. Some of these institutions and private investors became de facto TIMOs after they attracted other institutions to invest with them. The TIMOs in the United States have five distinguishable origins:

(1) insurance companies;
(2) banks;
(3) timber and timberland brokers, private timberland investors, and forestry consultants;
(4) investment firms and pension consultants; and
(5) non-US-based forestry investment management firms.

I now discuss these origins of institutional timberland investment and the pioneers and early practitioners of TIMOs. As we will see, many TIMO pioneers and early practitioners made entrepreneurial efforts in creating a new business in institutional timberland investment.

Insurance Companies

The earliest direct institutional timberland ownership in the United States was most likely Travelers Insurance Company (hereafter referred to as Travelers) in 1964. Founded in 1864, Travelers had been predominantly a general property and casualty insurer in the United States before it was purchased by Primerica in 1993 and became Travelers Group in 1995. Travelers Group merged with Citicorp to form Citigroup in 1998. After a few years, the insurance part of Citigroup again became an independent company called Travelers Company. Travelers as a prototype TIMO coincided with a period between 1953 and 1993 when it was purchased by Primerica, and what Travelers did during this period is of interest here.

Like other insurance companies, Travelers sought to earn a positive "spread" between the rate of return on its investments and the rate of return in credits to its policy holders. In its real estate investment department, Travelers had an agricultural division responsible for investment in agricultural properties, which sometimes had forests. Travelers had agricultural loans and bought timber bonds before the Second World War. But it was not until after 1953—when the Federal Reserve Board amended its regulations to allow financial institutions to lend money on timberlands—that Travelers made its earliest long-term loans purely on timberlands.[2]

In 1952 a young agricultural loan agent at Travelers, Dale Morrison, and his boss, Roger C. Wilkins, "stumbled" onto an opportunity to help St. Regis Company, a paper maker, purchase 140,000 acres of timberland along the Suwannee River in Florida. Subsequently, Morrison made several other timberland loans.[3] At that time, there was a lot of cutover and unmanaged forestland in the southeastern United States. Some forward-looking private landowners started to invest in forest regeneration and management to feed the demand of a growing forest industry. In particular, pulp and paper mills that had moved to the US South since the 1920s wanted to purchase or lease more forestlands to supply timber to their mills. By 1966, G. A. Fletcher, senior vice president of Travelers, estimated that all insurance companies collectively had more than $100 million invested in loans on timberlands in the South.[4]

Travelers' timberland loans were fixed-income assets that often required sufficient collateral and smooth income streams for repayment, but timber was initially perceived within Travelers as something that was hard to use as collateral and that could not generate a smooth income stream. Moreover, in Morrison's words, the landowners could "steal" the timber; their neighbors could mess it up; and because there was no fire insurance at the time, timber could be lost in a fire. Most importantly, Travelers did not know how to value timberlands, how to figure out timber growth, or how much to allow borrowers to cut timber to pay back their loans. It also had to deal with the possibility of poor forest management. The worst case was unclear boundaries for some timber tracts. Because of its lack of forestry expertise, Travelers lost some business opportunities, one of which was a proposed land purchase and lease deal brought out by the National Turpentine & Pulpwood Corporation in Jacksonville, Florida, in 1955.

Apparently, the National Turpentine & Pulpwood Corporation was trying to build a pulp mill in Jacksonville. In the 1950s, owning some timberlands was a must before a firm could raise money from debt markets to build a mill. Thus, National Turpentine & Pulpwood needed to fill in a "certificate of need" indicating that it had sufficient timberlands to supply wood for its proposed mill and thus could borrow money in financial markets. Guy Wesley, vice president of National Turpentine & Pulpwood, approached Morrison, trying to get Travelers to buy timberlands and lease them back to his firm. Travelers declined, citing the company policy of "devoting its investment entirely to mortgage and not to equity investments," even though Wesley's proposal was "a very attractive proposition."[5]

Shortly afterward, Morrison convinced the corporate leaders at Travelers that the company needed a forester to do forestry-related mortgage business with both forest industry firms and nonindustrial landowners. So, in 1958, Travelers hired Charles "Charlie" F. Raper, a 1954 forestry graduate from

North Carolina State University, as the first forester in the company and perhaps in the whole insurance industry in the United States. Born in Decatur, Georgia, in 1930, Charlie Raper was sent by his family to North Carolina State University to study forestry because his family owned some four thousand acres of forestland in Georgia. After receiving his forestry degree, he enlisted the US Army for a few years before joining Travelers.

In Morrison's words, Charlie Raper was "something else" throughout his whole career at Travelers. He was first stationed in Macon, Georgia, as a field representative and then rose to the rank of assistant regional manager, and field supervisor. In 1969, he moved to Travelers' headquarters in Hartford, Connecticut, as assistant secretary (1969–1973) and then secretary (1973–1982) of the agricultural division. In 1974, Raper received an MBA from the University of Connecticut. In 1982, he completed a law degree and was promoted to second vice president and head of the research and development division in the real estate department. Raper retired from Travelers and was the George W. Peake Jr. Professor at Auburn University from 1988 until his sudden death in 2000.

In 1964, Bowater Company, a UK-based firm that had paper mills in the southeastern United States, approached Raper for a deal similar to the one proposed by National Turpentine & Pulpwood nine years earlier. Bowater had issued some corporate bonds in previous years to finance its investment in mills, and the agreement on issuing the bonds stipulated that any additional timberlands it might buy afterward would be used as collateral for the bonds. So, Bowater had a hard time getting loans to buy timberland because no mortgage lenders were willing to be subordinated to the senior lender (or be "second in line") as far as collateral was concerned.

Bowater wanted Travelers to purchase some timberlands and then lease them back to Bowater. This time, Travelers agreed. Travelers formed a subsidiary called the Highway Land Company. Bowater would get its foresters to talk to willing private forest landowners, mostly farmers, to secure an option to buy their timberlands. After due diligence, the Highway Land Company would then buy the timberlands and lease the timber-growing rights on these lands back to Bowater. The deal was struck in the Southern Forestry Conference organized by the Forest Farmers (now the Forest Landowners Association) at the Grove Park Inn in Asheville, North Carolina, on May 7 and 8, 1964. The lease agreement was for twenty years and the realized real rate of return for Travelers, net of inflation, was about 4 percent when the lease agreement expired in the mid-1980s. For Travelers, it was a triple-net

lease—net of real estate taxes, management/maintenance, and insurance—
of some eighteen thousand acres of timberland.

In this way, Travelers became the first insurance company that had an
equity position in, and directly owned, timberlands in the southern United
States around 1964. In the next ten years, Travelers and Bowater made three
more similar deals, called Bowater II, Bowater III, and Bowater IV, respectively.
By the mid-1970s, Travelers ended up owning some one hundred thousand
acres of timberland in the South.[6] This was only a prototype of institutional
timberland ownership because one of the main goals of Travelers was still to
seek "fixed income" in this kind of land "buy-and-lease" deal.

Perhaps more importantly, Travelers had great success in its business
of timberland loans to nonindustrial private forest owners: it never experienced a default on its timberland loans. As Raper said in the late 1990s,
"Travelers has not lost a penny in timber."[7] On the other hand, Travelers was
not so fortunate in its agricultural loan business. In fact, it suffered big losses
in its agricultural loans in the late 1980s and early 1990s, which was one reason it became a takeover target and was eventually purchased by Primerica.

Based on his forestry knowledge and especially his experience with
timberland loans, Raper developed an investment proposition that timberland investments had low risk and could generate adequate returns. He
then advocated that Travelers should own more timberlands. In addition,
Travelers should invest some of its clients' money—mostly from pension
funds—in timberlands in a fiduciary fashion after the Employee Retirement
Income Security Act (ERISA) became law in 1974.[8]

Travelers did both. In 1979, it first started to own timberlands directly
as a form of investment for itself, and by 1984, it had acquired about 50,000
acres in Florida, Georgia, and Oregon. It also had two pooled accounts for
pension funds, completely separate from its own timberland investment
account. The first was an agricultural real estate fund, "Separate R," which
had some 3,800 acres of timberland in Georgia and North Carolina. The second, the "Separate Account T" fund, was a closed-end pooled account that
was invested exclusively in timberlands. Travelers marketed the "Separate T"
account in 1981 and raised $30 million when it closed in March 1984.[9] So, in
the early 1980s, Travelers was a "triple threat" in timberland investment—
timberland loans, direct timberland investing within its own portfolio, and
timberland accounts for institutional clients with fiduciary responsibilities.
The latter accounts make Travelers the very first large TIMO that we know
of today.

As a result, many who were in the TIMO business in the 1980s and 1990s considered Charles F. Raper the "pioneer of institutional timberland investment." Unfortunately, Travelers had to sell its own timberlands in 1993–1994 to cover its losses in agricultural and real estate loans and subsequently completely did away with its direct timberland ownership and fiduciary timberland accounts. Had it continued to actively market its timberland investment business in the late 1980s, Travelers could have been a dominant TIMO. Charley Tarver, the founder of Forest Investment Associates Inc. in Atlanta and another pioneer of the TIMO business, said, "Travelers could have dominated the TIMO business as it was so far ahead of everyone else in the early and middle 1980s."[10]

* * *

The exit of Travelers from timberland investment in the late 1980s and early 1990s coincided with the emergence of the John Hancock Insurance Company (Hancock) as a dominant player in the TIMO business for most of the next three-plus decades. Like Travelers, Hancock also had an agricultural and timberland loan business for many decades, and the earliest document or record of Hancock engaging in the agricultural mortgage loan business was in 1905.[11] All mutual life insurance companies that are owned by their policy holders need to invest in long-term assets that match their long-term liabilities. Operationally, they take the premiums or policies from their holders, make various investments, seek to have sufficient returns/reserves to meet their long-term liabilities, and make a profit. All these companies have a long-term perspective, but it is hard for them to lend to a company for thirty years when a ten-year duration may be the maximum for investment-grade corporate bonds. Land is different, and land investments are secure as long as the assets are protected.

Hancock had no forester in its Boston headquarters before Richard "Rick" N. Smith filled this void in 1980. Rick graduated with a forestry degree from the University of Maine in 1979 and was initially hired to manage, deal with, or dispose of a couple of timber properties in the northeastern United States that were in default (from loans) and were later repackaged and included in Hancock's first fiduciary agricultural real estate fund (ACRE Fund) in 1981.

After some study, Smith, like other pioneers such as Charlie Raper and Charley Tarver, also realized that timberlands were good investment assets. He subsequently recommended that Hancock own more timberlands instead of trying to get rid of them. So, in 1984, Hancock started a timberland-only commingled fund called ForesTree I. By all accounts, these two funds were small by today's standards: the ACRE Fund had $20 million and consisted of mostly agricultural properties, and after nearly two years of intensive marketing, the ForesTree I Fund had only $16 million when it closed in 1985, short of its $20 million goal.

Nonetheless, Hancock's marketing efforts in 1984 and 1985 paid off when the Ohio Teachers Retirement System and GTE Pension Funds each invested $25 million through Hancock (now Hancock Timber Resource Group, or HTRG; see below) in 1986. Since these investment amounts were large at that time, these two investors asked for separate accounts. Thus, the GTE Pension Fund was the first private pension fund and the Ohio Teachers Retirement System was the first public pension fund in the United States that had separate accounts, with Hancock as the timberland investment adviser and manager. At the same time that GTE gave $25 million to HTRG, it also gave $15 million to Forest Investment Associates Inc. (FIA) in Atlanta (see the next subsection).

What made Hancock thrive as a TIMO was the establishment of a consortium, or a business alliance called Hancock Timber Resource Group (HTRG), in the summer of 1985, even though the concept might have been conceived and discussed among its main players—such as Rick Smith of Hancock, Duncan Campbell of The Campbell Group, and Edward Givhan Jr. of Resource Management Service—a few months earlier. Like some other pioneers and early practitioners in the TIMO business, Smith saw potential for rising pension fund ownership of timberlands and thought that Hancock could capitalize on it, just as Travelers and Equitable Insurance (see below) had done.

Smith subsequently wrote a memorandum to Richard P. Troy, vice president of investment marketing at Hancock, on January 28, 1985, which was his concept paper. In the memorandum, Smith outlined the objectives and structure of HTRG and pointed out that Hancock needed to build a strategic alliance with forest management companies because Hancock did not have adequate local forestry expertise around the country.[12] This concept paper was praised by John Steel, president of Hancock, and the plan was put into action.

Soon, Hancock set up HTRG as a strategic alliance with The Campbell Group in Portland, Oregon, and Resource Management Service Inc. in Birmingham, Alabama. The two business partners covered two of the most important timber-producing regions in the country, the Pacific Northwest and the South. In 1990, the alliance added a fourth member, Wagner Woodlands & Co. in Lyme, New Hampshire, to cover its operations in the North and Northeast.

Smith's concept paper anticipated the request for proposals (RFP) for timberland investment advisers—RFP No. 85-31, issued on August 8, 1985—by the California Public Employees' Retirement System (CalPERS), the largest public pension fund in the country. CalPERS had already invested in real estate prior to this action. However, a change in California law (Proposition 21) on June 5, 1984, made pension fund managers in CalPERS think more broadly about real estate and about including timberlands, even though this law did not directly mention real estate or timberlands but only eliminated restrictions on the amount and category of stocks and assets that could be purchased by CalPERS funds. Prior to the enactment of Proposition 21, public pension funds in California could invest only up to 25 percent (but no more) of their portfolio in stocks, and these stocks had to be so-called blue-chip stocks (with company size and dividend requirements), and the rest was mostly in bonds. Proposition 21 gave CalPERS flexibility in determining its investment strategy.[13]

The decision by CalPERS to invest in timberlands was a lengthy process. Even before the passage of Proposition 21, CalPERS had hired Pension Realty Advisors of San Francisco, California, to evaluate investment in timberlands in response to the possible change in the California law. In a report on behalf of Pension Realty Advisors, William L. Ramseyer and Claudia Roberts Pitas wrote that as of December 1983, just under $100 million had been directed by pension funds to timber investments, which was minuscule relative to the $20 billion pension real estate equities that existed in the country. They further stated that some timber investments provided economies of scale and potential for good real returns without incurring many of the negative characteristics associated with nonurban real estate investment, because "timber appears to lack labor intensiveness, major weather and erosion problems, the need for quick management decision in response to weather changes, and also minor international competition and political interference relative to agriculture."

Ramseyer and Pitas further stated:

The best investments are made when circumstances allow assets to be available at favorable prices. . . . *Some parcels are probably for sale at prices that are attractive to the tax-exempt institutional investors given the real income stream that can be generated*[14] (emphasis added).

With this positive recommendation, the investment committee of the CalPERS Board of Trustees held seminars on timberland investment. On April 17, 1985, the investment committee informed the full CalPERS board that it was considering going through an RFP process for hiring timberland advisers through which the board would evaluate timberland investments.

On July 18, 1985, when the investment committee asked the full board to proceed with a possible RFP, the board was divided. Eventually, after lengthy debates and against strong opposition, the board voted six to four in favor of this motion: "The Committee recommends for approval of the RFP to be issued for the purpose of soliciting proposals from parties interested in becoming Board advisers on timberland investment."[15]

Hancock and a few other firms responded to CalPERS's RFP No. 85-31, but Travelers did not. The reason Travelers did not respond was that CalPERS's RFP stipulated that with a sixty-day notice, it could fire the timberland investment adviser. The lawyers at Travelers thought this could be a potential problem because it would be difficult to build a large management team and disband it in sixty days.[16] The lawyers at Hancock, at the persuasion of Rick Smith and others, agreed that it was not a problem.[17] In the end, CalPERS chose HTRG and Wagner Southern (see below) as the two finalists, or the winning bidders.[18]

Smith was instrumental in getting CalPERS's timberland investment business, and in the words of Jake Petrosino, then chair of the CalPERS board, "Hancock was more than accommodating" to CalPERS.[19] Through contacts with institutional investors, Smith was aware that CalPERS might be interested in investing in timberlands in late 1984. In fact, in marketing Hancock's ForesTree I Fund, he had gone to Sacramento, California, and lobbied CalPERS to invest in timberlands. Thus, as early as January 28, 1985, three months before the CalPERS board even discussed the possibility of hiring timberland investment advisers, Smith had clearly presented his vision in the aforementioned memorandum, aiming at getting the future timberland investment business from CalPERS:

Through CalPERS John Hancock has an excellent opportunity to make a critical advancement in its mission to be a bold and innovative force in the investment service market place. . . . Our success at attracting this business can be assured if we take a unified, quality approach in the structure of our proposal. Specifically, we must convince CalPERS that we have the depth and quality of personnel working as a cohesive, efficient unit. We must clearly show them that we have the ability to create and respond efficiently to "windows of opportunity" throughout the various "wood baskets". Because of all the factors surrounding this situation, *we should combine forces with regional timber investment groups to bolster our credibility and greatly increase our chances of success. The two groups which would add the greatest synergy to a Hancock Structure are Wagner Woodlands and Campbell Group.* An organization combining the fiduciary and investment strengths of Hancock with specialized timber investment skills at a local level of Wagner and Campbell, would perfectly fit CalPERS. *I am convinced we would sweep away the competition from the field*[20] (emphasis added).

In the same memo, Smith outlined CalPERS's situation and provided reasons behind the Hancock–regional partner strategy. CalPERS was looking for a cohesive organization able to execute its commitments. Yet despite its fiduciary experience with ACRE and the ForesTree I Fund, as well as its timberland mortgage lending business and its long involvement in forest industry firms in bond and equity issues, Hancock had only two field foresters and a part-time forest economist. Thus, to win the competition with potential competitors such as Travelers, Equitable, and specialized "boutiques," Rick Smith proposed that "Hancock form an alliance with two groups having highly specialized timber investment expertise at local levels—Wagner Woodlands/Wagner Southern Forest Investment, Inc. and Campbell Group."

He went on to propose the roles of Hancock and its partners in investment and marketing as well as compensation. There would be synergistic strength because regional groups exhibited the traits most highly sought by pension plans, such as local expertise and efficiency in management action, and Hancock fit the role of prudent adviser/fiduciary with its long experience in the timber investment market.

When approached by Hancock, Wagner Woodlands/Wagner Southern expressed the desire to stay independent. Hancock then started to look for another firm to serve as its southern arm. In the end, a prominent forestry consulting firm, Resource Management Service Inc., was chosen to play the role. In the spring of 1986, HTRG submitted its proposed timberland investment program for CalPERS.

On September 10, 1986, the investment committee of the CalPERS board approved the hiring of two timberland investment managers—HTRG and Forest Investment Associates Inc. (FIA, which evolved from Wagner Southern Forest Investments; see the next subsection)— allocated no more than $200 million to its timberland investment program, and authorized the staff to negotiate with these two managers on the terms outlined by its staff members. On February 11, 1987, the investment committee approved two separate fee structure and compensation packages for HTRG and FIA. The initial plan was to allocate $100 million to Hancock and $50 million to FIA, with another $50 million to be allocated once a satisfactory review was conducted.[21]

However, on June 16, 1987, the investment committee withdrew its approval of FIA as an adviser to CalPERS based on recommendation from its staff. The staff insisted that FIA provide CalPERS with an errors and omissions insurance policy, which would cost FIA about $20,000 annually, but FIA declined.[22]

So, CalPERS appointed HTRG as its only timberland investment adviser and gave it $150 million—first $100 million and then $50 million more after FIA was disqualified—to invest in timberlands in July 1987.[23] As noted, HTRG had $25 million each in earlier commitments from the Ohio Teachers Retirement System and GTE. Suddenly, in 1987, HTRG had more than $200 million invested in or committed to invest in timberlands. It surpassed Travelers (and Equitable Life Insurance; see below) in terms of total timberland assets under management and has since been a dominant player in the TIMO business.

Banks

First National Bank of Atlanta (FNBA) was the first bank to set up two pure timberland funds—one for tax-exempt institutional investors and the other for taxable individuals and institutions—in 1981. These were the first and perhaps only open-end timberland funds prior to 2015. Charley Tarver was instrumental in setting up these funds. He, along with two other individuals, founded FIA a few years later, which has since become one of the largest and longest-lived TIMOs in the United States.

After getting his forestry degree from Auburn University in 1968, Tarver enlisted in the US Air Force. He served during the Vietnam War and, as a captain, flew a tanker in the last US Air Force combat mission in Vietnam in August 1973. When he got out the air force in April 1974, the country was in a recession. He could not find a forestry or airline pilot job. He ended up working at the Citizens and Southern National Bank in Atlanta, which liked to hire well-trained and well-disciplined former military officers.

At that time, one of Tarver's friends, Jim Montgomery, worked for the Southern Forest Institute, an educational and public relations organization for the forest industry. The Southern Forest Institute was a branch of the

American Forest Institute, which had previously been called American Forest Products Industries, a trade promotion subsidiary of the National Lumber Manufacturers Association, and which later merged with the American Forest and Paper Association (AF&PA). AF&PA's predecessor started the American Tree Farm Program in 1941, a program promoting the establishment and management of forests on privately owned lands that continues today. By the late 1970s, this program had been promoted for thirty-plus years. With the support of major forest industry firms and federal and state governments, many farmers and other private landowners planted trees on their lands and registered their forests with the program. But many of these nonindustrial private landowners did not know how much their forests were worth, how to manage their forests, or what the timber markets would be when it was time for them to sell timber. On the other hand, forest industry firms supported the American Tree Farm Program as a means to ensure that a sustainable supply of timber would be available to support their regional investments in manufacturing facilities.

Most banks at that time had a trust department and still do now. Some of the trustees in these banks owned timberlands enrolled in the American Tree Farm Program. However, the managers of these timberlands in banks did not typically manage the timberlands actively because they were not foresters and did not know how to manage forests or when to sell timber. Jim Montgomery, as a promoter of sound forestry practices in general and the Tree Farm Program in particular, called all the major banks in Atlanta on his own and asked them to consider hiring a forester to manage the forests under their trusteeship. First National Bank of Atlanta (FNBA) agreed. Furthermore, FNBA wanted to hire not just a forester, but a forester who knew the banking business and its culture. Montgomery immediately recommended Charley Tarver. FNBA offered a job to Tarver, who accepted it and started to work in the real estate division of FNBA's trust department in fall 1979.

In early 1980 the manager of the investment division of the trust department at FNBA requested a meeting with the staff of the real estate division. The investment division had many clients who were institutional investors. Because the portfolio of the investment division then consisted only of stocks and bonds, its performance had been miserable in the previous two to three years. He was certainly not alone getting this result (figs. 3.1 and 3.2).[24] He was looking for new investment products and was thinking about getting into real estate.

The conventional vehicles for investing in real estate then were office and residential buildings, shopping malls, and other rental properties. In the

Figure 3.1. Annual rates of return for US common stocks, long-term government bonds, short-term government Treasury bills, and real property, 1970–1982

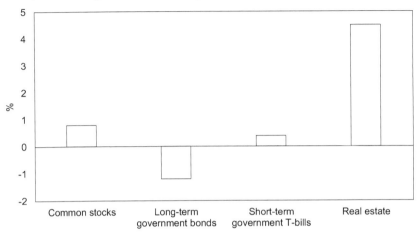

Source: Goldman Sachs (1983).

Figure 3.2. Timberland returns compared to other assets, 1960–1984

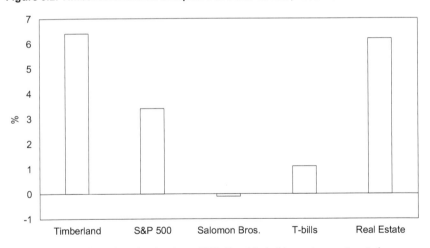

Source: Evaluation Associates Inc. brochure, 1985. The data in this graph cover twenty-four years (five years for real estate) ending on March 31, 1984. Salomon Bros. represents the average annual rate of return for the Salomon Bros. bond index. The timberland return was constructed by Evaluation Associates Inc. This graph was reprinted in Kirk (1985).

end, FNBA decided not to invest in these types of real estate for two reasons: it did not have the expertise, and others—especially insurance companies and other banks—had been doing it for a while and had a strong lead.

During the discussion, Tarver said that FNBA could invest in timberlands for these institutional investors. Nobody else in the room had heard about institutional timberland investment before.

Tarver was serious and dug into it. He studied and tried to find historical data on timberland investment. Very little information was available. So, Tarver on his own (and in collaboration with Charlie Raper) had to compile data and make some assumptions. Yet the more he studied, the more interesting it became to him and others at FNBA. He concluded that timberlands had desirable attributes and could be a new investment option for institutional and other investors. Tarver had the support of his boss, O. Medwin Welstad, who had written a term paper in graduate school at Michigan State University in the late 1970s that supported investing in southern rural lands.[25]

Eventually, the manager of the investment division liked the timberland idea and decided to give it a try using institutional funds over which FNBA had discretion to invest. FNBA subsequently informed its institutional clients of its intention to make this investment, set up the two accounts, and formally started to market timberland investment as a new product to its clients in 1981.

FNBA produced the earliest brochure for institutional timberland investment that I could find. Entitled "Timberland: A Growing Investment from the Ground Up," it described nearly all the attractive attributes of timberland investment as we know them today—namely, physical timber growth irrespective of market conditions, timber price appreciation because of ingrowth of timber and rising demand, potential increases in land value in the future, asset diversification, and hedge against inflation.

The ability of FNBA to use these funds over which it had discretionary power was critical. Otherwise, as Tarver said on February 16, 2015, "it might not have happened at all." At a minimum, it would have taken years for Tarver and his colleagues to raise the funds if FNBA had to market its timberland investment product and bring new money instead of merely informing and getting consent from its existing institutional and private clients.

Indeed, FNBA, Hancock, and other start-up TIMOs had difficulty raising money from institutional investors for timberland investment at the beginning. Tarver and his colleagues at FNBA were not able to attract much new money into these two funds between 1982 and 1983. Similarly, it took an intensive market effort for John Hancock Life Insurance to raise only $16 million in its ForesTree I Fund in 1984–1985, short of its initial target of $20 million. A more telling example was the US National Bank of Oregon, which was the largest bank in Oregon and had successfully managed timberlands in its trust department since 1959. In order to launch a timberland fund, it conducted a forestry investment fund study in 1982.[26] Yet it failed to raise merely $5 million in its Collective Timberland Trust Fund in 1985 after more than two years of marketing and fund-raising efforts, despite its experience and track record of managing forestlands on an Indian reserve and for other private forest landowners, a well-done feasibility study, and a brilliant brochure.[27]

With no prior experience or guidance from other companies, Tarver and his colleagues made both funds open-end because they thought it was fair to all investors. In December 1983, when Tarver left FNBA to start up a new joint venture called Wagner Southern Forest Investment Inc., the first fund had about $24 million in assets, and the second, tax-paying fund had about $4 million. Collectively, FNBA managed some twenty-six thousand acres of timberlands for its institutional and individual clients.

Wagner Southern Forest Investment Inc. (hereafter referred to as Wagner Southern) was a registered investment adviser on timberland investment in Atlanta, Georgia. It got its name from Wagner Woodlands & Co. Wagner Southern was backed financially by a UK firm named Booker McConnell Ltd. This relationship was initiated by George Hambrecht of Syrus Associates, a consultant for investment advisory firms in New York. Hambrecht first introduced Charley Tarver to Rodman Clark Rockefeller, a grandson of John Rockefeller.

Rodman Rockefeller had an agribusiness firm called International Basic Economy Corporation (IBEC) and was interested in starting a forestry consulting/investment business. By 1980, IBEC had merged with Booker McConnell Ltd. Rockefeller already knew Hank Swan, president of Wagner Woodlands. Because all three parties (Tarver, Wagner, and IBEC/Booker McConnell Ltd.) shared similar interests, they agreed to set up Wagner Southern as a new business venture. Booker McConnell Ltd. agreed to finance the operating budget of Wagner Southern for a few years, which was about $250,000 annually, covering the salary of three employees, travel expenses, and office rentals. In return, it would recover its investment with interest when the venture succeeded or exercise an option for controlling the interest of Wagner Southern.

On December 19, 1985, Tarver got a phone call from Hank Swan of Wagner Woodlands, with a message from Booker McConnell Ltd. saying that it would not provide any financial support to Wagner Southern after December 31, 1985. Apparently, after nearly two years of trial, the management at Booker McConnell Ltd. did not believe that the timberland investment business was going to work out and decided to walk away. Tarver and his partners had in effect lost their jobs a few days before Christmas 1985.

After two years of promoting their new business to numerous potential clients at FNBA and two more years at Wagner Southern, Tarver and his partners were convinced that they were on the verge of a successful venture. Tarver and Hank Swan of Wagner met, and both thought that they should continue to make their business succeed. However, they could not agree on which side should have a controlling interest in the firm. In the end, they decided to split. So, on May 9, 1986, Tarver, David Graham, and Kate Robie, all former employees of FNBA, founded Forest Investment Associates Inc. (FIA) in Atlanta, Georgia.

Soon after, FIA gained its first separate account of $15 million from GTE Company's Pension Fund, which, as noted earlier, also gave $25 million to HTRG for timberland investment. In early 1987, FIA secured another commitment of $10 million from the trust account of a wealthy family. Since then, FIA has thrived as a prominent TIMO.

FNBA was acquired by Wachovia Bank in 1985, which continued to be an important TIMO in the coming decades. As shown in chapter 5, Wachovia Bank grew a few new TIMOs on its own in the 1990s and 2000s.

Timber and Timberland Brokers, Private Timberland Investors, and Forestry Consultants

Many forestry companies—from timber and timberland brokers to forestry consultants—have handled timberland transactions and advised landowners on forest management on both a short-term and long-term basis. A few of these companies have gone one step further to become the managing partner of limited partnerships that own timberlands, thus assuming the management responsibility of other partners' investments in timberlands. When these limited partners include institutional investors, these companies become de facto TIMOs even though they continue to do their traditional brokerage and forest management work. One of these companies is Canal Industries.

* * *

Canal Industries, based in Conway, South Carolina, was once the largest wood dealer in the country and one of the largest private owners of timberland in five states in the southern United States between 1960 and 1990. Canal Industries was chartered on September 10, 1937, by E. Craig Wall Sr. and R. H. "Hutch" Gibson in a fifty-fifty partnership with a total investment of $2,000. Three decades later, it became a multimillion-dollar empire.

Canal Industries was managed by E. Craig Wall Sr., who was an astute and shrewd businessman. Over the next few decades, he developed an uncanny knowledge about land value, whether he was spending $10 or $40,000 per acre. "Buying into the path of progress" was what he called it, as he bought or traded for many pieces of timberland that had potential for development.[28]

Canal Industries owned more than three hundred thousand acres of timberland in the Southeast in 1969 when the son, E. Craig Wall Jr., a graduate of Davidson College and Harvard Business School, took over as the president and chief operating officer. E. Craig Wall Sr., who would serve on the Richmond Board of the Federal Reserve Bank in the 1970s, maintained the title of chair and CEO until his death in 1985.[29]

Craig Wall Sr. and Jr. believed in doing business in partnerships, even after they could afford to carry deals alone. After Craig Wall Jr. took over, he saw that Canal Industries needed new partners to grow its business because it was "land rich" and "capital poor." Thus, in 1972, he and one of his associates, Walter Herbst, embarked on an eighteen-month journey, marketing timberland investment on Wall Street and to their business contacts, and essentially inviting others to invest in timberlands with Canal Industries in partnerships. It was a difficult process; as he recalled many years later to Charles Godfrey, former copresident of Canal Industries, "Nobody on Wall Street took our idea seriously."[30]

Nonetheless, two close family and business friends bought their idea. One was Hugh L. McColl Jr., chair of the board of the North Carolina National Bank Corporation (NCNB, now Bank of America) in Charlotte, North Carolina. The other was Joe L. Roby, chair of the investment firm Donaldson, Lufkin and Jenrette (DLJ) in New York. Both McColl and Roby accomplished many business deals in their respective careers and would become influential figures in the US banking industry in the later part of the twentieth century. Hugh McColl Jr. merged NCNB with a California-based Bank of America and served as the chair of the new Bank of America before retiring in 2002. Joe Roby became the chair of Credit Suisse First Boston LLC, a subsidiary of Credit Suisse Group, before his retirement in 2004.

In 1973, Hugh McColl Jr., then with NCNB, decided to put up some of the bank's money and partner with Canal Industries, and DLJ helped secure two additional institutional clients: the Eastern Airlines Pilots Pension Fund and the AT&T Pension Fund. Although it is unclear how much money NCNB and the AT&T Pension Fund contributed, the Eastern Airlines Pilots Pension Fund contributed $5 million, or some 5 percent of its total pension fund, which was big in timberland investment at the time. This investment in timberlands from Eastern Airlines was the result of a series of events brought about by Charles "Chuck" Dyer.[31]

Chuck Dyer graduated from Bowdoin College in 1965 and Harvard Business School in 1967. Dyer served as a US naval aviator from 1958 to 1972 and was a pilot for Eastern Airlines between 1967 and 1991. In 1970, at age thirty-two, he was assigned to a five-person pilots' pension and benefit negotiation team at Eastern Airlines because he had a business degree. This team would negotiate for more than one year with Eastern Airlines management, under the leadership of senior vice president Frank Bowman, who had been a US astronaut and commander of Apollo 8, the first manned spacecraft to orbit the moon, earlier in his career. After the negotiation concluded in December 1971, Dyer was named the director of the pension plan at Eastern Airlines because his predecessor had taken a job with the pilot union in Washington, DC.

One day in the spring of 1972, Dyer went to see his pension fund manager, US Trust in New York. He was "thrown out" on the street by a security guard for not having an appointment. In an interview with me on February 25, 2015, he said with a laugh, "I was well dressed, but I might have looked too young." It was then that he decided to change the pension fund manager of the Eastern Airlines Pilot Union. With the approval of his union, he fired US Trust by the end of 1972.

In the process of searching for new managers, he called two of his prep school classmates and fellow Harvard Business School graduates, Charles D. Ellis and Lewis Kresch, who happened to be working at DLJ at that time. Charles D. Ellis was another accomplished business leader who later went on to establish Greenwich Associates and wrote several best-selling books, including *Winning the Loser's Game* (McGraw Hill Education, 1998, 2002, 2010, 2013), *The Partnership: The Making of Goldman Sachs* (Penguin Press, 2008), and *What It Takes: The Secrets of Great Professional Firms* (John Wiley, 2013). As Chuck Dyer wanted to diversify a small portion of his investment into real estate, Ellis introduced him to the real estate division at DLJ. With the support and approval of his boss, Joe Roby, Kresch took Dyer to see the two Walls of Canal Industries in 1973.

The exact motivation and expectation of the two other partnering institutions—NCNB and AT&T Pension Fund—are unclear. In Chuck Dyer's words, NCNB had a very aggressive real estate division in its trust department. The closeness of Hugh McColl with the two Walls must also have contributed to NCNB's decision to invest in timberland in the early 1970s.

Eventually, these three institutional investors—Eastern Airlines Pilots Pension Fund, NCNB, and AT&T Pension Fund—set up a company called Atlantic Land Corporation, and Canal Industries set up a subsidiary called Canal Land Fund Inc., both on February 15, 1974. These two companies would become a partnership called Canal Land Company, which was renamed Canal Land Limited Partnership in 1986.[32] Canal Land Fund Inc. served as the general partner of Canal Land Company and took over the management responsibility of existing lands, new land purchases, and disposals. Canal Land Fund Inc. also initially contributed some timberlands to the partnership, which were appraised by Harry Schroeder of Joseph Blake & Company of Dallas, Texas. Other than cash (timber) income, all property under management would be appraised annually. However, because the business was set up as a partnership, there was an exit strategy for its partners.[33] In 1986, the partnership was reorganized.

This partnership or joint venture was run well until Eastern Airlines went bankrupt in 1992 and Craig Wall Jr. died prematurely at age fifty-nine in 1996. When the assets of Canal Land Limited Partnership were sold off in early 2000, there were three investors on the account: Eastern Airlines Pilots Pension Fund (which owned about 48 percent), North Carolina National Bank Pension Trust (5 percent), and Marine Crew (47 percent). Apparently, AT&T must have pulled out sometime earlier and Marine Crew came in.

In retrospect, the two Walls at Canal Industries were also pioneers of the TIMO industry long before the term "TIMO" was coined. They attracted institutions to invest in timberlands using a limited partnership as the conduit. They did not use the term "commingled fund," but their partnership was no different from the partnership arrangements used by many TIMOs today. They first attracted private individuals' money to timberland investment and then purposefully and successfully marketed timberland investment as well as their own expertise, experience, and success to institutional investors.

Just like Canal Industries, Georgia Timberlands Inc., another timberland landowner, broker, and timber dealer, based in Macon, Georgia, also had a successful record in timberland investment and nearly became a TIMO. Georgia Timberlands used Travelers for timberland loans and ended up managing some timberlands for Travelers. As stated, Georgia Timberlands owned two hundred thousand acres of timberland in Alabama, Georgia, and North Carolina in the early 1980s. Unfortunately, Georgia Timberlands was bought out by a local financier in 1986 because of its high debt and mismanagement. Many other companies around the country have managed family-owned forestlands for a long time, such as Seven Island Land Company, which has managed the Pingree family timberlands in northern and western Maine since 1964; the Port Blakely Companies, which have managed the Eddy family timberlands in the state of Washington since the 1920s; and Prentiss & Carlisle, which have managed land for a variety of clients, including several long-standing timberland-owning families since the 1920s. It is possible that some of these companies might have invested institutional monies in timberland and/or managed properties for institutions in the 1970s and 1980s. Because they did not actively seek institutional clients to invest in timberlands, they are not considered the pioneering firms of the TIMO business or even TIMOs as we know them today. On the other hand, Lone Rock Resources in Oregon is now considered a TIMO after it transitioned from focusing only on the Sohn family forestlands in the 2000s to now including institutional investment partners.

* * *

Like Canal Industries, a few private timberland investors and forestry consultants had similar timberland partnerships first with private investors and then with institutional investors. One of them, Wagner Woodlands & Co. in Lyme, New Hampshire, was owned by Frederick "Fred" E. Wagner before his death in 1981.

Wagner was born in 1916 and grew up in a wealthy family in Milwaukee, Wisconsin. He graduated with a bachelor's degree in liberal arts from Dartmouth College in Hanover, New Hampshire, in 1934 and a master's degree from Harvard Business School in 1948.[34] At Dartmouth, he enjoyed off-campus life and was friendly to a farmer who owned a seven-hundred-acre farm with some timberlands. He bought the land with his family money in his last year of college and planned to live there. When the United States joined the Second World War, he enlisted in the US Navy. After the war, he participated in a family business venture in Wisconsin. He "retired" at age thirty-nine and returned to the Hanover area of New Hampshire in 1955.

Initially, Wagner was interested in agriculture and bought a dairy farm on the Connecticut River, along with some 1,400 acres of timberland at two dollars per acre, "almost for free." In the following years, he was fascinated by the tree farm business and started to buy timberlands for investment purposes for his family. He received help and advice from the New England

Forestry Foundation as well as Allison "Al" Cateron, who was a former district forester with the US Forest Service in Boston, and Allen Page, a consulting forester at Black Diamond Forestry in Belchertown, Massachusetts. By 1960, he had accumulated some ten thousand acres of timberland and hired Al Cateron as the full-time chief forester of his family timberland business.[35]

Wagner was an avid reader and quick learner. He read forestry books and learned practical forest management practices from foresters. When he was developing his timberland investment idea in the early 1960s, he used a model of a hypothetical one thousand acres of white pine forest with certain growth rates and ingrowth and price appreciation to show that timberland investment would have handsome returns. He showed this result to his family members and prospective investors. But at the beginning of his timberland investment, he used mostly his own or his family's money. He was wealthy, but not extremely wealthy. He had a very sharp mind and was a shrewd businessman looking for any good timber-growing land with timber. One of his employees, Bob Berti, remembered that the first property he bought for Wagner Woodlands after he was hired by Fred Wagner cost about fifty dollars per acre in the early 1970s—yet the timber alone was worth more than fifty dollars per acre. Timberlands in New Hampshire were very inexpensive then. A tract of timberland with trees eight to ten inches in diameter would sell for only a few dollars per acre. Even silviculturists thought it was a good investment.[36]

On September 1, 1966, Wagner set up a limited partnership company and named it Wagner Woodlands & Co. and served as the president and general partner. He also set up various partnerships with his schoolmates and Dartmouth and Harvard alumni, focusing on timberland investment. His business partners were mostly wealthy individuals and families such as the John Rockefeller and Robert Winthrop families. As noted earlier, Rodman Clark Rockefeller, grandson of John Rockefeller, had connected Hank Swan of Wagner Woodlands and Charley Tarver in setting up Wagner Southern in Atlanta in 1984.

The company registration certificate that Wagner Woodlands & Co. filed with the Secretary of State Office of New Hampshire in 1966 stated:

> The business purpose of this partnership is to acquire, develop, operate and dispose of, for the partnership account and for the account of others, all manner of real estate including, without limitation, *commercial timberland, Christmas tree plantations, wildling Christmas*

tree stands, tree nurseries, gravel pits, mineral rights, water rights and recreational real estate; *to provide professional forestry advice to others; to perform all manner of forestry operations for others*, and to engage in such other businesses as are incidental to or useful in conducting any of the foregoing (emphasis added).

At the beginning in 1966, all the partners in his firms were Wagner family members. To convince these partners to invest, Wagner would offer them 4–6 percent interest if they invested a sizable amount of money. Wagner succeeded and had a good track record. Gradually other private investors joined.

His first nonfamily partner was William Taylor, a Dartmouth alumnus who worked for an oil company. Together they bought some five hundred acres of timberland on a fifty-fifty basis in 1966. The total investment was about $50,000 at about one hundred dollars per acre with timber. Wagner's share in the ensuing limited partnerships would gradually be reduced to one-third in the later 1960s, and eventually down to 1 percent in 1974. He then expanded to Maine and other states in the Northeast and to Quebec in Canada. His vision was to buy large tracts at a discount, ideally five hundred acres and larger.[37]

In 1972, the certificate of the Wagner Woodlands & Co. limited partnership was amended to include another limited partner outside the family, Thomas H. Choate. Choate contributed $120,000 and was awarded a 5 percent interest in the company. In other words, Wagner Woodlands & Co. was worth about $2.4 million ($120,000 divided by 5 percent) in that year.

In 1974, Wagner had learned from John Wallis of the University of New Hampshire that returns on timberland investment rose faster than inflation. He hired Ken Super to conduct a one-year study, which validated his intuition. This study gave him added confidence in timberland investment. His vision in the mid-1970s was that forest growth came from three components: biological growth (4 percent per year for intensively managed white pine), ingrowth or product movement (4 percent), and price appreciation (2–3 percent), which together would generate a compound rate of return of 10–12 percent. This was exactly the same investment formula developed by Georgia Timberlands in the 1970s.[38] The only difference was that Georgia Timberlands' rate of return was 12–14 percent or more because trees grow 5–6 percent annually in Georgia, much faster than in the northeastern United States.

In 1976, Wagner went to court to fight an inside takeover attempt by David M. Roby, a young and ambitious lawyer whom he had hired a few years ago to handle his expanding partnership business. Roby eventually went his own way and founded The Lyme Timber Company LP, also in Lyme, New Hampshire, in 1977. At the beginning, The Lyme Timber Company had the same business model as Wagner—namely, limited partnerships with private individuals on timberland investment. The Lyme Timber Company had four principals and some twenty limited partners between 1976 and 1993, all of whom were private individuals. In 1993, it secured its first timberland investment from an institutional investor and became a TIMO as well.[39]

In 1977, Wagner hired Henry "Hank" Swan, a forester, from the agricultural loan division of Hancock. Swan advised Wagner to go after institutional investors, in addition to the traditional individual private investors. Wagner secured an institutional investment from the New England Life Insurance Company in 1977. In 1983, Hancock requested that Wagner serve as the manager of properties in its ACRE Fund.

As his company grew, Wagner was under a lot of stress because of business and family reasons. He committed suicide on November 1, 1981, and was remembered as a fine gentleman with vision and integrity. He gave his employees freedom to practice sound, sustainable forestry and expected integrity and performance in return. Just like the two Walls in Canal Industries, Wagner was one of the pioneers who discovered the trick of timberland investment, put his own money into play and established a solid investment track record, attracted private and institutional investors to invest in timberlands, and eventually led his firm to grow into a TIMO before the term was coined a decade or more later.

At the time of Fred Wagner's death, Wagner Woodlands & Co. owned or managed some 225,000 acres of timberland in the northeastern United States and eastern Canada. The company would choose Hank Swan as its new president, and all the limited partners agreed to make Swan the new general partner. Most investors stayed with the company afterward. Swan later set up Wagner Southern with Charley Tarver, which preceded Forest Investment Associates.

Hank Swan was also instrumental in getting more TIMO business for the company. In the mid-1980s, Wagner managed some $50 million in timberland for wealthy private individuals in the United States, Europe, and a Catholic church as its clients in the timberland business.[40] As noted earlier, Wagner Woodlands became a partner with HTRG in 1990 and changed its name to Wagner Forest Management Ltd. in 1993. After the HTRG partnership was dissolved, initially in 1997 and finally in 2004, Wagner Forest Management

Ltd. continued as a TIMO and with its forestry consulting business. It does not do much marketing or fund-raising but rather focuses on property management for private and institutional investors.

* * *

Other notable early practitioners from the management or supply side of institutional timberland investment are Duncan Campbell, founder of The Campbell Group (now Campbell Global); Ed Givhan (Edward Holmes Givhan Jr.), former president of Resource Management Service; and James A. Rinehart, then associate professor at the University of California at Berkeley. Here we use Campbell and Rinehart to illustrate their thought processes and actions.

Campbell got his accounting and law degrees from the University of Oregon in 1973. He was thus a lawyer and CPA. Prior to forming The Campbell Group, he was a tax manager at Arthur Anderson. In this capacity, he served multiple public and private forest products companies. His areas of expertise included the tax treatment of timberland as well as the unique aspects of investment in forest products for both taxable and nontaxable investors. It was at Arthur Anderson that he developed his timberland investment ideas.

Campbell saw that among all the Fortune 500 companies, forest products companies had some of the lowest returns in the early 1980s. Yet these companies acquired a lot of timberlands to ensure a continuous supply of timber to their mills. Because of high interest rates in the early 1980s, Campbell reasoned:

> During the 1980s, the forest products industry, particularly in small and medium companies, would find a need to build a timber supply without sufficient internal capital. Consequently, this limits their availability for timber financing. Recognizing this, The Campbell Group intends to profitably emerge as the leader in generating capital investments in timberland, not only to fill the forest products industry timber supply needs, but to meet the financial and tax goals of individual investors.[41]

So, Campbell started to question why forest products companies bought and held timberlands. He thought it was possible for someone else to buy the timberlands from these companies, manage them, supply timber to these companies, and get the required rate of return, while the companies whose cost of capital was higher than these rates of return could use the proceeds from the sales to invest in other production activities.[42]

On October 9, 1984, at a timberland investment conference titled "Timberland Market Place for Buyers, Sellers, Investors and Their Advisors" in Portland Oregon, organized by the Duke University Center for Forestry Investment, Campbell stated:

> I have no doubt in my mind that over the next two decades, other than by some major forest products companies such as Weyerhaeuser, there will be a major shift of ownership. That shift of ownership will go from the forest products companies to investment groups; domestic, international, institutional, and wealthy individuals. And the reason is this: timber, let alone land, cannot generate a rate of return that is going to meet internal corporate goals. However, it will generate a rate of return that is compatible for investors, especially for institutional investors.[43]

On the same day and at the same conference, James A. Rinehart delivered a paper with an almost identical proposition: the costs of industrial timberland ownership could outweigh the benefits to forest products companies. Rinehart reasoned that the cost of capital for forest products companies in the early 1980s was in the middle double digits, and the returns on timberland were less than 10 percent. Thus, companies would be better off selling their timberlands to tax-exempt institutional investors who could bear a low but stable rate of return.[44]

One year later, Rinehart published his paper, titled "Institutional Investment in U.S. Timberlands," in the *Forest Products Journal*, and it has since been credited as the first scientific article proposing the separation of timberlands from the manufacturing business of forest products firms. Rinehart would become the director of portfolio strategy for HTRG between 1990 and 1992, and he formed the GATX Capital Timberland Investment Group as a TIMO to make equity investments in timberland in 1998, where he worked until 2005. Rinehart had his own timberland investment advising firm, R&A Investment Forestry, before his retirement at the end of 2019.

Both Campbell and Rinehart, as well as a few others, reached the same conclusion by watching what was happening in the market as Sir James Goldsmith and other corporate raiders were buying forest products firms and selling them in parts (see the next chapter). The proposition of Campbell and Rinehart is illustrated in figure 3.3, where the hurdle rate, or required rate of return for a forest products company, is R. The forest products company has two business segments: timberlands and manufacturing plants. If the required rate of return for new investment in manufacturing plants (R_P) surpasses its overall hurdle rate (R), and that for timberlands (R_T) does not, the forest products company should sell its timberlands to tax-exempt institutions, which may have a lower required rate of return because they are tax exempt and often look for diversification benefits from their investments. The minor difference between them is that Campbell saw that small and medium forest products firms would have a hard time finding internal capital resources to hold or buy additional timberlands to meet their needs for timber, while Rinehart saw that the opportunity costs of holding on to industrial timberlands were already too high and called for a divestment.

Figure 3.3. A hypothetical investment choice for a forest products company that owns both timberland and manufacturing plants

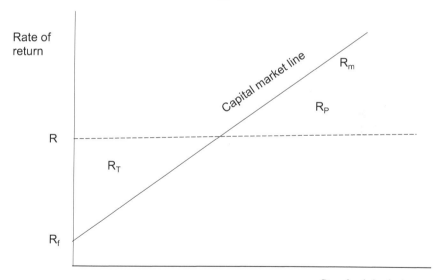

Note: The capital market line links the risk-free rate of return (R_f) and the returns of a broadly defined investment portfolio (R_m). The horizontal axis represents the total risk of asset returns, measured by their respective standard deviations. R represents the overall required rate of returns for the forest products firm. R_P represents the hypothetical rate of return from manufacturing plants, and R_T stands for the hypothetical return from timberland. When R is greater than R_T, the company is better off selling its timberlands. Also, because R_T is above the $R_f R_m$ line, meaning that timberland investment has an adequate return compared to its risk, there should be investors interested in buying timberlands. Note that R_P could be to the left of the capital market line ($R_f R_m$) as well.

In theory, it is not necessarily true that timberland returns are too low to meet forest products companies' hurdle rates for the separation of timberland from manufacturing plants. It could be the other way around—the manufacturing plants are too inefficient to generate sufficient returns on the capital employed. For one thing, an efficient capital market can price various components of a forest products firm whose cost of capital is the weighted average of the cost of capital for the various lines of business it holds. Furthermore, many forest products companies, especially those with a lot of pulp and paper mills, used cheap timber to prop up inefficient manufacturing facilities. Timberlands were generally a cost center, not a profit center, so their managers were judged by how cheaply they could provide wood to a mill.[45] Third, as noted, GAAP completely ignored the fundamental production process underlying timberland investments, so much so that industrial timberlands were even considered "dead assets" in the 1980s.

In any event, based on the premise of likely changes in existing industrial timberland ownership, Campbell converted his firm, Campbell Forest Resources Inc., which was set up on April 30, 1982, to The Campbell Group on April 4, 1984. Again, its business focus was to raise capital primarily from individual investors, pension plans, and/or trusts for the acquisition and management of timber resources to fill the need for capital of the forest products industry as well as reach the financial and tax goals of investors.[46]

As part of the HTRG business alliance after 1985, The Campbell Group helped HTRG grow to over $2.5 billion in timberland assets under management and garnered nearly 50 percent of the market share in institutional timberland investment in the United States when the alliance was dissolved in 1997. As early advocates and practitioners of institutional timberland investment, Campbell, Rinehart, and Givhan made a big contribution to the development of the TIMO business.

Investment Firms and Pension Consultants

Eric Oddleifson graduated from Harvard College in 1956 and Harvard Business School in 1963. He started his career as a consultant with Arthur D. Little in Cambridge, Massachusetts. In 1975, he found a job with B. F. Saul Company in Chevy Chase, Maryland, where he first researched and then implemented a farm investment program in southwestern Georgia. After Frank Saul sold off the farm business in 1980, Oddleifson returned to Boston, seeking other investment opportunities.

An alternative capsule autobiography that Oddleifson wrote between 1975 and 1982 stated:

> I was asked to head up an investment program to acquire and operate land for various agricultural crops. My research had suggested that the southeast U.S. offered opportunities for double cropping under irrigation, and for the next five years I acquired and managed an intensively operated farmland portfolio, including drip irrigated pecan groves. Improving return by applying technology to natural resources became my business objective.
>
> It was here that I learned to question financial projections, and the assumptions on which they are based. Corn yields failed to materialize, as I had mistakenly counted unproductive lowland as able to produce 200 bushels of corn. Corn prices behaved respectably at first, moving upwards from $2.50 per bushel. My projections showed corn prices reaching $7.00 per bushel in the early 1990's. Quite the opposite happened. They went into reverse, and now trade at slightly over $2.00 in current dollars.
>
> I also cringe at my clearing of 15-year old pine stands on class one agricultural lands, in order to plant irrigated soybeans. *That mistake cost me 10 percent real returns from tree growth alone* (emphasis added). I did learn, belatedly, the folly of my ways from a timber consultant, Eley C. Frazer, of F&W Forestry in Albany, Georgia, who showed me how to project timber growth and yields as the basis of an understandable investment product.
>
> This, plus my look-back, in 1980, at institutional investment returns the previous 20 years (zero percent real) convinced me that timber investing, with its six to eight percent real returns on a conservative basis would be of interest to the institutional market.[47]
>
> After an abortive solo attempt to establish my own asset management company, and down to my last meal, I walked into the Boston Company in the summer of 1981, looking to both sell the timber investment idea and get a job. I was successful at both, with Ed (C. Edward) Broom (who was in charge of new ventures for the company at the time) co-founding with me Resource Investments in March of 1982.[48]

Ed Broom was then the senior vice president at the Boston Company, an investment management holding company. He was responsible for institutional business, including identifying new business and investment opportunities for institutions, mostly pension funds. Oddleifson asked Broom to put timberlands in an individual institution's investment portfolio. Broom initially did not think that timberland investment would work.

After about a six-month dialogue including consultation with Travelers and Hancock, Broom agreed to give timberland investment a try. In March 1982, Oddleifson joined Broom and they set up a subsidiary called Boston Company Resource Investments. Because they thought that no single institution would have the amount of money needed—or more precisely the interest in providing the amount of money needed—to buy large tracts of timberland, they started to market for a pooled fund.

Broom left the Boston Company in April 1983, shortly after it was acquired by American Express, and became the executive vice president of Evaluation Associates Inc. (EAI), an investment consulting and money management firm. Broom then asked the Boston Company whether he could take Boston Company Resource Investments with him. His request was granted because Boston Company Resource Investments had spent a lot of money studying and marketing the timber investment business, but it had not had any clients or revenues. Thus, Oddleifson and the former Boston Company Resource Investments created Resource Investments Inc. (RII) as a subsidiary of EAI in 1984.

While they were at EAI, Broom and Oddleifson went to the Equitable Assurance Society of the United States (Equitable) in 1984–1985 to set up their first pooled Timberfund. They had good contacts and potential clients in Equitable, and they were not sure about any structuring options other than insurance company group annuity contacts they could employ without triggering unrelated business taxable income (UBTI) for institutional investors. At that time, to avoid triggering UBTI, other timberland funds were created by either insurance companies such as Travelers and Hancock through group annuity contracts, or banks such as First National Bank of Atlanta through IRS (US Internal Revenue Service) private letter rulings.[49] Therefore, their first Timberfund, established in 1984, was called the Equitable Timberfund, of which EAI was the fund manager. Equitable marketed its Timberfund aggressively and had some immediate success, as its first Timberfund had $50 million with two segments—$35 million and $15 million, respectively.[50]

In 1985, Mutual of New York (MONY) acquired EAI and thus owned RII. However, RII was treated as an independent firm as it started to pay its employees' salaries from management fees it charged on institutional timberland assets under its management. In 1989, Oddleifson, Broom, and Steve Hurley bought RII from MONY for about $1 million and made it an independent company. In 1995, RII was sold to Union Bank of Switzerland (UBS), and the "Inc." became "International" so that the acronym RII stayed the same. As was the case with Hancock and FNBA/Wachovia Bank, several TIMOs later emerged from RII, following the retirement of Broom and Oddleifson in 1997.

RII was then under the asset management division within UBS and Oddleifson was the managing director. Because both Oddleifson and Broom came from the demand (pension management and advising) side, not the supply (forestry) side of institutional timberland investment, they too are considered early practitioners of institutional timberland investment in the United States.

Non-US-Based Institutional Investors

Although US timberlands might have been owned directly by foreign institutions many decades ago, the earliest direct, non-US-based institutional timberland investment in the United States that I could trace was in the late 1970s.

In 1979, the British Coal Board pension fund came to the United States with a $50 million investment in timberlands. The person who made this happen was Alexander Fell, a forester with the Economic Forestry Group in Scotland. This was a direct investment of $25 million through Billy Humphries, owner of Forest Resource Consultants in Macon, Georgia, and another $25 million through a timberland buy-and-lease agreement with International Paper Company. The latter was similar to the agreements Travelers made with Bowater in the 1960s and 1970s. The British Coal Board pension fund acquired some seventy-five thousand acres of timberland in Georgia and Louisiana. The investments performed poorly because they were made at the top of a market cycle. Timber prices did not return to their 1979 peak level, upon which the transaction was priced in the US South, until the mid-1990s. This, plus a change in leadership of the British Coal Board, led to the sale of these US properties in 1987.[51]

Similar to the United States, the United Kingdom experienced double-digit inflation after the OPEC oil embargo in 1973. Individual and institutional investors were looking for assets that could provide protection against inflation. Alexander Fell learned that the lumber price index in the United States was indeed rising faster than the rate of inflation. He also saw that forests grew about 3 percent annually on average in the United States, which was higher than in the United Kingdom, and that forests grew even faster in the southern United States. On the other hand, the United Kingdom did not have a lot of forests in which to invest. The British Coal Board had even proposed buying public forests in the United Kingdom in the 1970s.[52] After its proposal was rejected by the British government, it started to seek timberland investment opportunities overseas and made investments in the United States.[53]

Another British forestry firm that facilitated institutional timberland investment in the United States was Fountain Forestry. In 1953, a Scottish landowner and member of a Lloyd's insurance company said that his company could offer better fire insurance rates for young plantations than those being offered by other insurance companies. He further suggested that some of the directors and underwriting names might like to invest in forestry because by planting trees, they could offset the cost against their other incomes and thus reduce the very high taxes they were paying at the time. Some of the directors indeed bought some forest properties. To manage these properties, these landowners together established Fountain Forestry on November 12, 1957.

Soon, Fountain Forestry started to manage timberlands for other investors, including some institutional investors. As Fountain Forestry grew, with the encouragement of its long-standing clients, it considered developing its managerial model in an overseas market. In the summer of 1980, Fountain Forestry bought three properties in the eastern United States for individual clients, and in 1981, it set up a US operation called Fountain Forestry Inc. in Boston. Over the next few years, the Plessey Pension Fund from the United Kingdom invested directly in a number of forestland properties, for which Fountain Forestry became the manager. Fountain Forestry was later moved to New Hampshire and continued to manage timberland throughout the United States for individuals and institutions, irrespective of their nationality.[54] In 2016, Fountain Forestry was bought out by a prominent forestry consulting firm, F&W Forestry, whose headquarters is in Macon, Georgia.

The other foreign pioneering TIMO is Brookfield Asset Management Inc., a Canadian asset management company listed on the Toronto and New York Stock Exchanges. Brookfield started to own and manage timberlands through investment in forest products firms that owned both timberland and forest products manufacturing facilities in Canada, the United States, Brazil, and Europe. It then started to invest in just timberlands. As of December 2016, it managed over $4 billion, or 1.5 million hectares (3.7 million acres) of timberland assets in the United States, Canada, and Brazil, but it had only $800 million in timberland assets under management in 2020, mostly in Brazil.

There were also foreign investors in US timberlands from Japan, China, the Middle East, and Europe in the 1970s and early 1980s. Some of these investments were direct, meaning that the investors managed their timberlands directly. Some were associated with ownership of forest products manufacturing facilities. The sources of investments were corporations, governments (sovereign funds), individuals, and institutions.

Finally, note that there are many non-US-based institutional investors in global timberland markets other than the United States. For example, Forestal Caja Bancaria is an investment of the Uruguayan Banking Pension, a social security institution founded on May 14, 1925. The start of its forestry activity dates back to 1964, when it purchased 14,826 acres in Paysandú, Uruguay, to plant pine and eucalyptus trees. In the following years, small adjoining fields were added. In 1992, it acquired 14,085 acres in Durazno and 11,861 in Paysandú. Now it owns a total of 44,478 acres of forestlands and employs more than three hundred direct or indirect collaborators.[55] Similarly, Caja Notarial (Notary Pension Fund), a nonstate public agency, has also had forestry investments since the 1960s, and Uruguayan government pension funds have been investing in timberland there since 1994.

The biggest TIMO outside the United States before 2010 was Société Forestière de la Caisse des Dépôts (Forestry Company of Deposit Funds) in France. Founded in 1966, it has managed forests for major institutional investors as well as private owners in France ever since. It is now one of the largest forest management companies in Europe. The company had about 307,000 hectares of forest under management in France, with an estimated asset value of more than €2.0 billion in 2020. Similar to Canal Industries and Wagner Forest Management Ltd. in the United States and to Fountain Forestry in the United Kingdom, this French company attracted these investments first from private individuals, and then institutions. More on TIMOs outside the United States is included in chapter 7.

Some Failed Efforts

I must point out that the creation of sufficient demand for institutional timberland investment and of the TIMO business was not easy and that the choice of operational modality and management style was important. This is illustrated by some failed efforts by investment banks and companies on Wall Street.

At roughly the same time that a few TIMO pioneers gained some success in their new business, some investment banks and large investment firms on Wall Street also started to consider organizing and advising timberland investments. In the early 1970s, Oppenheimer & Co. had a closed-end timberland limited partnership for private investors and used Georgia Timberlands as one of its managers. This partnership was terminated in the early 1980s, possibly because its investments were made when timber prices were high. Similarly, Merrill Lynch's farm and timberland division did some timberland business, mostly as a timberland broker, between 1980 and 1985. In 1982, E. F. Hutton, a New York–based investment and brokerage company, started to invest directly in timberlands from its own account. It formed a subsidiary called Hutton Timber Resource Corporation, which offered private investors timberland investment opportunities through the limited partnership vehicle. By 1986, it had attracted over $38 million and bought approximately forty thousand acres of forestland in Georgia and Florida. The publicly offered Hutton Southern Timber Partners I and II included Hutton Timber Resource Corporation as general partner and some 5,500 limited partners.[56]

None of these investment banks or large investment companies lasted long in the timberland investment business. Perhaps their management fees or hurdle rates were too high for a fledgling timberland investment business in the high-inflation era of the 1970s and 1980s. Perhaps timberland investment advising required specialized forestry knowledge. It is also possible that they did not fully understand the cyclical nature of the timber and timberland business.

Thus, only the very few companies that had long experience in timberland investment via debt or debt and equity gained the trust of institutional investors and became successful TIMOs in the early stage of institutional timberland investment. These companies included insurance companies (Travelers and Hancock), banks (FNBA, Wachovia), private timberland investors (Canal Industries), and forestry consultants (Wagner Woodlands). Of course, there were also FIA, an "outgrowth" of FNBA; The Campbell Group; and non-US-based firms that handled direct investments by foreign pension funds in the United States. The only exception was Resource Investments Inc., which came from the demand side.

The Emergence of Institutional Timberland Investment and TIMOs in the United States: A Summary

Institutions—especially insurance companies and banks—got into equity investment in timberland between the 1960s and 1970s as they learned from their earlier debt investment in, and trusteeship of, private forestry. Because they understood the risk-return characteristics of timberland investment, they not only invested in timberland on their own but also wanted to manage timberland investments for other institutions for a fee. At roughly the same time, a few private timberland investors such as Canal Industries and Wagner Woodlands, which had secured good returns from their own equity investments in timberland, had also purposefully invited institutional investors to join them. TIMOs started to emerge in the United States.

In the 1970s and early 1980s, when inflation rates in the United States were in double digits, timberland started to attract more private and institutional investors in the United States and overseas as an inflation hedge and portfolio diversifier. Better yet, timberland investment had adequate returns that were comparable to and even better than those of stocks and bonds in the preceding decade. Consequently, Travelers, Hancock, Canal Industries, Georgia Timberlands, Wagner Woodlands, and a few others bought and managed timberlands for private and institutional investors.

Figure 3.4. Major events in the development of TIMOs in the US, 1964–1986

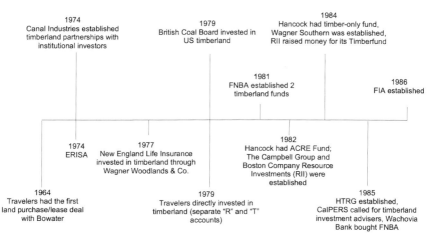

Thus, TIMOs as a business in the United States emerged from multiple sources (fig. 3.4). Like other successful entrepreneurs, the pioneers of the TIMO business all had foresight, vision, and determination. They saw what was ahead in the timberland investment business and acted. In the process, they overcame difficulty and adversity with their tenacity and created a demand that had largely not existed before. TIMOs started mostly in the early 1980s, so those who worked in TIMOs at that time and before are all innovative pioneers, the most notable being Charlie Raper, Craig Wall Jr., Fred Wagner, Charley Tarver, and Rick Smith. Hank Swan, Eric Oddleifson, Ed Broom, Duncan Campbell, Edward Givhan, and Jim Rinehart are some of the most noticeable early practitioners and advocates.

So far, we have covered the emergence of institutional timberland investment and a few pioneers of TIMOs in the United States. But who were the sellers of timberlands, and why did they want to sell in the 1980s and afterward? The next chapter covers the early sellers and buyers of timberlands and their motivations.

CHAPTER 4
The Sellers and Buyers of Industrial Timberlands in the 1980s–1990s

As some institutional investors began to look for timberlands through newly emerged TIMOs or on their own, they found plenty of timberlands on the market across the country in the mid-1980s. Most of these timberlands were large tracts and came from vertically integrated forest products firms; thus they could qualify as "investment-grade" timberlands that were productive and located in active timber markets, had appropriate tract sizes, and were offered at attractive prices. This chapter is about the supply and demand conditions for these "investment-grade" industrial timberlands in the 1980s. On the supply side, a few market, regulatory, and institutional factors made some firms change their timberland ownership policy and abandon the idea of pursuing a high timber self-sufficiency rate. A few firms started to sell all their timberlands. On the demand side, a different set of market and regulatory factors made more institutional investors start to explore the viability of timberlands as an investment vehicle. A few of these investors were convinced that there was an unprecedented opportunity in timberland investment, partly because they valued timberlands differently from industrial owners. As a result, an arbitrage opportunity arose. At the end of this chapter, I show some of the market inefficiencies in timberland markets in the 1980s that institutional investors, through TIMOs or on their own, were then able to exploit.

Three key concepts create the analytical framework of this chapter: managerial control versus shareholder value; cost centers versus profit centers; and various tax issues—including corporate capital gains versus ordinary income, double taxation versus single taxation, and no taxes on passive income for institutional investors. The presentation and discussion of these concepts are interweaved throughout this chapter.

The Sellers: Why Divest Industrial Timberlands?

As noted earlier, until the early 1980s most forest products companies had taken a corporate attitude of acquiring more timberlands. They saw that timberlands were a prerequisite of manufacturing and therefore were building large timberland bases around their manufacturing facilities. Weyerhaeuser Company had assembled six hundred thousand acres in eastern North

Carolina just to support its two paper mills in Plymouth and New Bern. In the later 1970s when it announced a plan to build a new paper mill in Columbus, Mississippi, it first assembled six hundred thousand acres in western Alabama and eastern Mississippi.[1] Container Corporation of America, a midsized forest products company based in Chicago, Illinois, had a goal of 60 percent self-sufficiency for its fiber in the early 1980s, doubling the self-sufficiency rate of the industry at the time.[2]

As these companies purchased lands from nonindustrial private landowners and as new mills were built, lands owned by one company might be closer to another company's mills. Thus, when oil prices and freight costs escalated in the 1970s, competing companies began to exchange lands among themselves. Weyerhaeuser thus made several transactions with Union Camp, Champion International, and Federal Paper Board for twenty thousand to thirty thousand acres in each transaction.

Then came the economic recession in 1980–1982 and a prolonged crisis in the forest products industry. The severe recession was attributed to macroeconomic policy that resulted in record high interest rates and a collapse of US housing markets, and it ended in late 1982 when the US economy started to grow again. However, the crisis in the forestry industry lasted much longer. Timber prices, especially federal timber prices in the US Pacific Northwest, increased dramatically in the 1970s and then collapsed after 1979 (fig. 2.2).

Western forest products companies that bought federal timber at much higher prices before the recession secured a bail-out plan through political maneuvering. The bail-out plan—the Federal Timber Contract Payment Modification Act of 1984—allowed these companies to leave their timber contracts with the federal government by simply paying a buyout fee in whole or in increments. This relief thus helped some small and sizable firms restructure and head off bankruptcy.[3]

But this bailout was not enough to save several large forest products firms. A series of takeover attempts took place. English financier Sir James Goldsmith snapped up two large forest products firms—Diamond International in 1982 and Crown Zellerbach in 1985—and profited by trying to buy two other forest products firms—St. Regis Corporation and Continental Corporation. Newspaper magnate Rupert Murdoch, with assistance from the owner of an international construction company in Kansas, bought the Continental Can Corporation and made it private. As Preston E. Kirk pointed out, "The wolves, having gotten the scent of weak balance sheets, however, have not been chased away from the doors of those companies."[4]

What Sir James Goldsmith did after these two takeovers gave a bigger shock to forest products firms than the takeovers themselves, and it truly challenged the prevailing view of industrial timberland ownership. After gaining control of Diamond International and Crown Zellerbach, Goldsmith immediately split the timber management department from the manufacturing department in each company and asked each department to maximize its own returns independent of the other—the land management side was to sell its trees or logs wherever it could achieve the highest prices, and the manufacturing side was to procure its raw materials wherever it could find the cheapest delivered costs. It was the first time that a forest products company was viewing its own land as an asset and not as an extension of manufacturing. Sir James Goldsmith said, "In breaking up an old conglomerate, you're not killing companies, you're not buying them to shut them down and sell their assets. You are simply killing off the bureaucracy and letting the companies inside the conglomerate free."[5]

As a consequence, selling industrial timberlands to improve their balance sheets and treating timberland operations as a profit center independent from their manufacturing business were seen by forest products firms as two viable options to fend off potential corporate raiders. Previously, most forest products companies, especially paper companies, had organized their timberlands as a cost center—the timberland managers were tasked to deliver the needed quantity of pulpwood to the log yards at the lowest possible cost, including both company timber and purchased wood.

James River Corporation, a paper company based in Richmond, Virginia, that bought some paper assets from Sir James Goldsmith after he gained control of Crown Zellerbach, was one of the first forest products companies that followed the new design by Sir James Goldsmith—treating its timberlands as a profit center.[6] Although this action did not solve the undervaluation of timberland on the company's balance sheet because of accounting rules, it nevertheless demonstrated that its manufacturing business could be profitable without timberlands.

Similarly, after ITT Rayonier Corporation did a spin-off and formed Rayonier Timberlands Limited Partnership in 1985, Rayonier started to break the supply link between its timberlands and manufacturing facilities, treating both as separate profit centers. Soon, very little or no Rayonier wood was going to its sawmills. That quite naturally led to Rayonier being able to sell its sawmills, which it did. This auctioning of timber to the highest bidder helped establish the nonintegrated operating model that Rayonier continues to use today.

Several other companies took the other route: they tried to monetize their timberlands and fend off potential hostile takeovers. Before we turn to the monetization option, let us see how Sir James Goldsmith found his targets and implemented his takeover strategy and how forest products companies responded in the face of real and potential hostile takeover bids.

Hostile Takeovers and Undervaluation of Industrial Timberlands Based on Book Value

Between the 1930s and mid-1980s, the prevailing corporate governance philosophy was to expand and diversify horizontally and vertically at the corporate level, because diversification was thought to lead to corporate profitability and stability. Horizontal diversification occurs when a firm develops or acquires new products that are different from its core business or technology. Vertical diversification or vertical integration occurs when a firm goes back to previous stages of its production cycle, such as self-supply of raw materials in a backward integration, or moves forward to subsequent stages of the same cycle, such as distribution of its final products in a forward integration.[7]

Vertical integration to timberlands had been encouraged for forest products firms before the 1980s. Coincidentally, this period was labeled a period of managerial control in the corporate governance literature. Such managerial control was possible because corporate management had a lot of power to make management decisions such as diversification, shareholders were numerous and had limited influence over management, and legal restrictions limited the ability of shareholders to act collectively.[8] The corporate managers in this era might not have been as keenly interested in the profitability of their firms as in the firms' long-term stability and their own ability to rise within their respective internal corporate hierarchies. They had large managerial autonomy and were effectively insulated from individual shareholders.

In the early 1980s, a new form of corporate control was rapidly developing, forcing corporate managers to pay closer attention to their companies' balance sheets so as not to find their companies the potential targets of hostile takeovers. At that time, a merger movement was washing over the United States as antitrust measures were weakened under the Reagan administration, and with the rise of mutual funds, institutional investors and a reinvigorated shareholder group demanded higher returns on their investments. A new concept of corporate control—shareholder value—emerged, asserting that corporations were for the purpose of preserving and increasing shareholder value and did not need to become a conglomerate or diversify into various unrelated businesses, since diversification could be achieved at the shareholder level. This movement represented a historical transformation of corporate America to the shareholder value concept of corporate governance, where shareholder value was seen as the first and only objective of a firm.

In the forest products industry, Sir James Goldsmith single-handedly implemented this shareholder value concept of corporate governance by starting and carrying out a list of hostile takeover activities. Goldsmith was an Anglo-French financier who began his career in various food manufacturing industries in the United Kingdom. His business partners described his philosophy succinctly as the belief that "the sum of the parts of most conglomerates was worth a great deal more than the whole."[9] For forest products firms, the crown jewel was their timberlands, which might have been purchased a long time ago and thus have a very low book value because of inflation over time and depletion.[10] But the market values of these timberlands were much higher, and a 1982 study by Morgan Stanley even stated that appraised timber values for twenty-two large forest products companies alone exceeded their market value by 89 percent.[11]

Historically, forest products companies treated their timberlands as an extension and insurance of forest products manufacturing. In a comprehensive study published in 1968 on timberland ownership by the US pulp and paper industry, Norman D. Hungerford found that security was the number one justification for timberland ownership and that these companies incurred high opportunity costs for this security. He pointed out:

Basic in the rationale of most companies (with respect to timberland ownership) is an overriding concern with security. . . . Although a favorable return on timberland investments is highly desirable, it is not the major factor governing a company's decision to own or not to own forest land. *The capital represented in company timberlands could most certainly provide greater returns in a number of alternative investments within a firm, at much lower risk. In this light, companies are presently experiencing major opportunity costs arising out of the ownership of timberlands.* There is little to suggest that wood obtained from company lands over that obtained on the open market provides any material cost savings; more likely the reverse is true. *From a strictly economic point of view, it is extremely difficult to rationalize the present extent of forest land ownership in the pulp and paper industry*[12] (emphasis added).

Because timberlands were not treated as a profit center, timberland managers were not in a position to bargain with mill managers. There were anecdotal cases in which sawmill and paper mill managers sometimes offered lower prices for timber coming from their own lands than for timber from open markets.

This low transfer price itself was not necessarily a bad thing. Every mill has a delivered cost curve. The highest-cost sources of wood are open market purchases of the last logs needed to keep the mill running, and arguably industrial timber is closer to the lowest-cost sources of wood, on average or at the margin. The problem was that when companies did not value their internal wood supply using market metrics, they did not account for all the costs of investment in planting, silviculture, management, and property taxes, or the time value of money in valuing the cost of wood from their owned timberlands, because of accounting rules.[13]

In essence, many large forest products firms contained assets on their balance sheets that were worth more than the net capital value of their stock in 1982. For example, and as cited by Kathleen K. Wiegner in a *Forbes* magazine article, Crown Zellerbach had an estimated $69 per share in its fair market timberland value alone, while its stock traded at $18 per share; International Paper had $73 per share in timberland and trees, while its stock was selling at $45 a share (table 4.1).

The financial conditions for some forest products firms in the early 1980s provided savvy investors with a prime opportunity to purchase a controlling

Table 4.1. Total assets and estimated timberland value of major forest products companies, 1982

Company	Total assets ($ million)	Timber estimated market value ($ million)	Timber value per share ($)	Recent stock price ($)
Weyerhaeuser	5,716	8,035	60	34.5
International Paper	5,544	3,607	73	45.0
Georgia-Pacific	5,060	2,781	25	25.5
Champion International	3,443	1,686	28	19.5
Boise Cascade	2,740	1,502	56	31.5
Crown Zellerbach	2,614	2,049	69	23.5

Source: Wiegner (1982)

share of these firms, to strip and sell off their assets in pieces, and to gain a large profit. This takeover strategy must also imply that forest products companies could proactively monetize their timberlands. Donald Brennan, who left International Paper as vice chair to head up Morgan Stanley's newly formed forest products group, stated, "There is a tremendous imbalance between the book value of timber and the market value. Some place between those numbers there is an opportunity to monetize that timber."[14] This monetization strategy will be discussed later in this chapter.

As noted earlier, Goldsmith implemented his takeover strategy on four forest products companies. Interestingly, St. Regis Corporation fended off Goldsmith's hostile takeover bid by asking Champion International to serve as a "white knight" and come to its rescue.[15] In order to defeat Goldsmith's takeover efforts, Crown Zellerbach was the first Fortune 500 firm to adopt an antitakeover measure that soon became widely known as the "poison pill," to no avail.[16]

Diamond International Corporation (Diamond), headquartered in San Francisco, was the first target of Goldsmith's debt-financed takeover strategy and is used here to illustrate Goldsmith's tactics. Diamond was established in the late nineteenth century and was the country's leading manufacturer of matches. By 1980, it had diversified into a number of unrelated product lines. The result was that even though it had a total sales revenue of $1.2 billion, it had only some $40 million profit, a meager profit margin of 3.3 percent in 1977.

In 1978, Ira Harris of Salomon Brothers brought Diamond to Goldsmith's attention. In poring over Diamond's balance sheets, Goldsmith's associates noticed that Diamond owned 1.6 million acres of timberland carried on the books at about $50 per acre. Thinking this must be a mistake, Goldsmith's associates asked Salomon Brothers to double-check. It was true. Diamond had purchased most of its timberlands around 1900, depleted the timber assets when it harvested original and mature timber, kept a low book value for these timberlands according to GAAP, and never revalued them, even though timber might have grown back on these timberlands.[17]

Subsequently Goldsmith quietly purchased about 5 percent of Diamond's shares in early 1979. In order to borrow money to finance his bid for Diamond, he went to Travelers to establish a value for the timberlands owned by Diamond. With a fee of $12.5 million, Travelers agreed to pay Goldsmith $250 million for the 1.6 million acres of timberland if and when Goldsmith got control of the timberlands and decided to sell them.

In essence, Travelers had written Goldsmith a put option on these timberlands.[18] Because the exercise price for this option was very low (at only $156 per acre, or three times book value), Travelers would profit whether Goldsmith exercised it or not.

Goldsmith was certainly not interested in selling these timberlands at that price after he gained control of Diamond. He used Travelers merely to establish a market value of the timberlands, which allowed him to borrow money from debt markets to purchase Diamond.

In March 1982, Goldsmith put together the financing for his Diamond bid and succeeded on November 1, when the shareholders voted overwhelmingly to accept his offer. By 1983, timber analysts were estimating the value of his timberlands alone at $723 million, which was about $500 per acre, or ten times the book value. In the end, *Fortune* magazine estimated that Goldsmith netted a $500 million profit and ranked the Diamond deal as "one of the most profitable financial events of the 1980s."[19] Interestingly, Sir James Goldsmith sold all other assets and kept timberlands. He said, "*Timberland in the USA is as good a long-term investment as you can find and you need to employ very few people. It was the jewel in the Diamond crown*"[20] (emphasis added).

Forest Products Companies' Initial Defense—Outright Sales or Master Limited Partnerships

Forest products companies that had already been under stress because of the recession and crisis in the forest sector were thus put on notice by Goldsmith, Robert Murdoch, and other "corporate raiders." Wall Street also noted the difference between the book value and market value of timberlands. In the aforementioned study in 1982, Morgan Stanley estimated that the timberlands alone would be worth more than the companies as a whole by a substantial margin.[21] Despite the assumptions and errors made in deriving these appraised values of industrial timberlands, it was clear that Wall Street was encouraging monetization of these timberlands in the early 1980s.

Thus, other than treating timberlands as a profit center independent of manufacturing operations, forest products companies had pressure to either sell at least some of their timberlands outright or find another way to monetize their timberland holdings. Some companies indeed put their timberlands up for sale. Since unloading their timberlands outright would require these firms to find a new class of timberland owners who could afford to pay for large tracts of timberland, this eventually led to the emergence of, or at least strengthened, institutional timberland ownership and TIMOs. Yet because the prevailing management philosophy was to hold on to timberlands, other forest products firms favored another means of monetizing their timberlands without selling them outright.

The advice from Wall Street was to set up timberland master limited partnerships (MLPs) if forest products companies did not want to sell their timberlands outright. The MLP structure was first utilized in 1981 by the oil and gas industry. Specifically, the term "master limited partnership" was coined after Apache Petroleum became the first firm to merge with several illiquid limited partnerships into a single "master" limited partnership and served as its general partner. This is the aggregation or "rolling up" approach of MLPs. The second method is the "spin-off," accompanied by the contribution of corporate assets and followed by the public offering of units (or shares). Most timberland MLPs took the spin-off approach.

Masonite Corporation formed the first public timberland MLP for its 466,000 acres of timberland and sawmills on July 28, 1982. International Paper Company (for 6.3 million acres on December 20, 1984), ITT Corporation (Rayonier for 1.2 million acres on October 3, 1985), and Pope and Talbot Incorporated (under the name Pope Resources for 82,000 acres on October 28, 1985) followed. Together, these companies spun off 8.1 million acres, which was about 13 percent of all industrial timberlands owned by forest products firms at the time.

Masonite and Pope distributed limited partnership units to their existing shareholders. On the other hand, International Paper and ITT Corporation (Rayonier) marketed a minority interest in their limited partnerships to the general public and thus continued to control these timberlands. Rayonier would become timberland REITs two decades later. In any event, the timberlands owned by institutions via TIMOs or timberland MLPs/REITs initially were mostly industrial timberlands.

The story of Masonite Corporation perhaps best illustrates the pressure on forest products firms that owned a lot of timberlands. Masonite was a conservative, closely held, publicly traded, midsized forest products firm. It had little debt, and management owned one-third of the company's common stocks. Masonite became a takeover target after its share prices dropped because of the 1980–1982 recession. There were two unsolicited hostile takeover attempts. One was from an investor in Canada. Another was from the Pritzker Group in Chicago, Illinois. In response to the first hostile takeover attempt, Masonite decided to increase its debt and change its bylaws by adding a poison pill clause to make itself less attractive. When the second hostile takeover attempt came not long afterward, it decided to self-liquidate most of its assets in five years.[22]

On August 26, 1982, the stockholders of Masonite Corporation approved a plan of partial liquidation. Under the plan, Masonite discontinued its southern and western lumber and woodlands businesses and transferred all assets and related liabilities of these businesses to a new, publicly traded Mississippi limited partnership called Timber Realization Company. It did so because its board of directors felt that this was the best way for stockholders to realize the value of the timberland properties. James Bearrows, who was associated with Timber Realization Company, stated:

> After careful study, Masonite decided that the timberlands were not necessary to its ongoing operations and that the stockholders would obtain more of their appreciated value by orderly disposition than by the company continuing to hold them. The limited partnership structure was thought to be a perfect vehicle for doing this.[23]

Eventually, the Timber Realization Company sold all of its 466,000 acres of timberland and ten sawmills in the South and California and ceased operations by 1987.

International Paper set up its timberland MLP (IP Timberlands Limited) differently. IP Timberlands Limited was a continuing partnership. It had two units, A and B. Public investors who purchased "A units" had the right to an income stream from International Paper's 6.3 million acres for fifteen years. More specifically, "A unit" holders received 95 percent of the net income until 2000. Beginning in 2000, 95 percent of the net income flowed primarily to holders of "B units," which were all held by International Paper itself, and the remaining 5 percent share went to holders of the "A units." In other words, International Paper monetized its timberlands largely by selling the right to net income associated with these timberlands for fifteen years. By the end of 1985, it had sold 16 percent of its "A units" to the public.

This was a purely defensive move—a poison pill—as it did not improve International Paper's earnings outlook or benefit current shareholders. How would this move have worked against a potential takeover? If a hostile bid had come up, International Paper could have spun off the partnership (by selling all remaining "A units" to the public and "B units" to existing shareholders) to make the company more expensive and more difficult to acquire. In this way, the timber and timberland would not have been available to the prospective buyer as collateral to finance a potential takeover. Furthermore, without a call on its extensive timber assets, the value of the standing International Paper mills would probably have been reduced.[24]

As noted earlier, Crown Zellerbach, too, wanted to establish an MLP for controlling and eventually liquidating its 1.6 million acres of timberland as a poison pill. However, Goldsmith was able to take over Crown Zellerbach before this timberland MLP could be launched.[25]

Forest Products Companies' Underperformance, Mergers and Acquisitions, and Heavy Debt

Despite resistance and mitigation efforts by executives and managers, the effect of the hostile takeover movement on forest products firms in the 1980s was profound. Over the course of the 1990s and early 2000s, a number of changes took place that again forced corporate management to increasingly adopt the shareholder value concept for their firms—so much so that the 1990s would become a decade in which shareholder interest replaced managerial control in forest products firms. And because the compensation and incentive packages offered to executives and managers were more aligned with shareholder value, the interests of shareholders became the interests of management.[26]

Let us look at the external environment faced by forest products firms in the 1990s and early 2000s. First, the bull market of the 1990s was an extraordinary period of growth in the US stock markets. Comparatively, the financial performance of the forest products sector was greatly lagging that of the broad markets, as shown in figure 4.1, for a twelve-year period from 1988 to 1999. During this period the S&P 500 increased by nearly five times, while the index of forest products increased by less than half that. Similarly, in the decade between 1995 and 2005, the rate of return for the forest industry was 6.2 percent, which was only about half the 12.1 percent in the S&P 500.[27] If we go back to the 1970s, forest products companies also underperformed the broader markets because they had to invest heavily in emission control for their paper mills to comply with the Clean Water Act, and because of overcapacity.

Second, rapid globalization in the 1990s brought increasing competition to US forest products firms, which was compounded by technological changes during this period. For example, information technology and the proliferation of computers led to a decline in paper consumption, especially newsprint and, to a lesser extent, printing and writing paper, in the United

Figure 4.1. Monthly stock price index for Weyerhaeuser, S&P 500, and S&P 500 Paper & Forest Products Index, 1988–1999

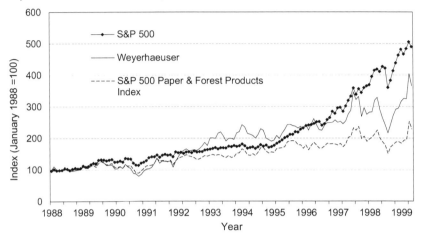

Data source: CRSP (Center for Research in Security Prices at Wharton Research Data Services). The whole paper and forest index underperformed the S&P 500 in the entire twelve-year period by a large margin. Note the overperformance of Weyerhaeuser stocks during the period of high timber prices in the early 1990s, and the underperformance during the recession in 1991 and the Asian financial crisis in 1998.

States. It did not help that some countries' governments and development banks were financing and even subsidizing the construction of pulp mills at that time.

Third, the decline in the public timber supply led to an increase in timber and timberland prices that greatly benefited large timber firms during this period and enhanced the appeal of timberland investment among institutional investors. As shown later in this chapter, the benchmark NCREIF Timberland Index had an average rate of return of 22.5 percent in its first ten years of existence between 1987 and 1996.

Under pressure for better financial performance, forest products firms used various tactics and strategies, such as increasing stock buybacks and dividend payments, mergers and acquisitions, restructuring, layoffs, and sales of manufacturing assets and timberlands. For our purpose, we note only that some stock buybacks and dividend payments as well as mergers and acquisitions were financed by debt. As figure 4.2 shows, there was a drastic increase in the debt-to-equity ratio in the US paper products sector in the early 1990s, and this ratio remained at high levels for the next decade, largely as a result of the merger activities that took place on an unprecedented scale of multibillions of dollars in the late 1990s and early 2000s (fig. 4.3). Partly because companies wanted to reduce their debt after megamergers and acquisitions, they collectively sold, or converted into REITs, forty million–plus acres of industrial timberlands between the 1990s and early 2000s.[28]

Double Taxation and Change in Capital Gains Tax for Industrial Timberland Owners

Another very significant factor on the supply side of industrial timberland sales was double taxation of forest products firms. Most forest products firms are Subchapter-C corporations that pay federal income tax on their earnings at the corporate level. In addition, when dividends are paid to shareholders, they pay individual income taxes on those dividends. Partnerships, MLPs, or REITs are pass-through or follow-through entities that do not pay corporate income tax; earnings are taxed only once at the ownership level. Thus, most forest products firms are tax inefficient insofar as their timberlands could be changed to timberland MLPs (and later, REITs). Corporate structure of timberland entities matters.

One thing that somewhat helped forest products firms reduce their tax burden on timber income was the capital gains tax treatment of timber income that had been in place since 1943. Capital gains tax regulations affect (1) the treatment of income associated with harvesting trees and (2) the income associated with selling timberland. The impact of capital gains tax on income from timberland sales can be completely eliminated through either timberland MLPs or a Reverse Morris Trust, to be discussed in chapter 6,

Figure 4.2. Debt-to-equity ratio in the US paper products sector, 1960–2004

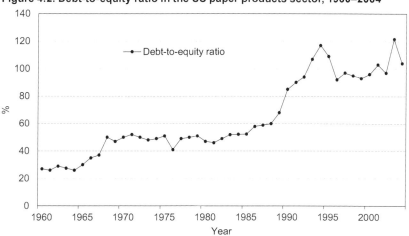

Source: "Pulp & Paper Global Fact & Price Book 2005–2006," in *Pulp and Paper,* ed. Will Miles (Bedford, MA: Paperloop, 2007), as cited in Gunnoe (2012).

Figure 4.3. Mergers and acquisitions in excess of $1 billion in the US forest products industry, 1995–2007

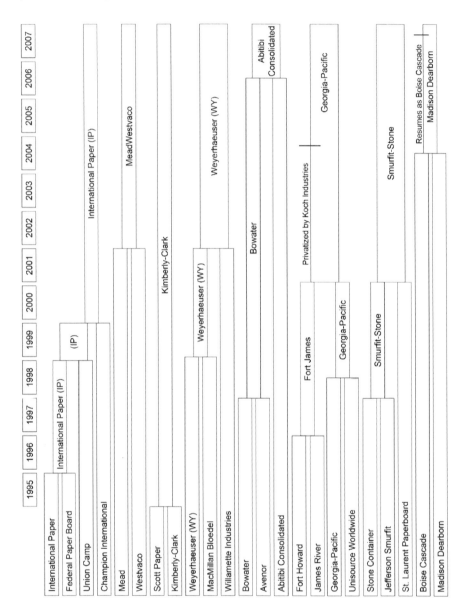

Sources: "Pulp & Paper North American Fact Book," in *Pulp and Paper*, ed. Greg Rudder (Bedford, MA: Paperloop, 2002), and "Pulp & Paper Global Fact & Price Book 2005–2006," in *Pulp and Paper*, ed. Will Miles (Bedford, MA: Paperloop, 2007), as cited in Gunnoe (2012).

and alleviated by the installment sales technique, to be discussed in the next subsection. Here we focus on capital gains tax on timber income.

Usually, an integrated company had to separate its income into income arising from manufacturing activities and income from timber sales. It paid capital gains tax on timber income and ordinary corporate income tax on income from manufacturing activities. The integrated company often did this via an internal transfer price for timber coming from its own lands, and the IRS had strict rules about how that price could be set. Since manufacturing income was taxed at a higher rate than timberland, in some cases the company might have enough operating income losses to fully offset all its timber capital gains. Thus, timberland provided a great tax shelter for integrated companies.[29]

In 1986, however, this capital gains tax treatment for timber income received by industrial owners was "eliminated." Technically it was not truly eliminated, but timber income generated from industrial timberlands was taxed at the same rate as other corporate income. All things being equal, industrial timberland owners were worse off after the 1986 tax reform. Until the very recent reduction in the corporate income tax rate to 21 percent (effective on January 1, 2018), roughly in line with the highest current capital gains tax rate of 20 percent, corporations had a taxation disadvantage relative to nonindustrial timberland owners, who paid mostly capital gains tax on timber income, and institutional timberland owners, who did not pay any taxes on passive income including dividends and land sales.

The elimination of capital gains tax for corporate timber income further tilted the balance from industrial timberland ownership toward institutional timberland ownership. As a report by Lehman Brothers in 2006 shows, the difference in timber incomes between a double-taxed C corporation and a tax-advantaged financial ownership such as a TIMO or REIT could be as high as 39 percent.[30] Although owning timberland may be advantageous and profitable, integrated forest products companies can by no means compete with tax-advantaged firms with such a high margin.[31]

Although US corporate tax rates have been set at a flat rate of 21 percent and capital gains tax at 20 percent in the highest bracket since January 1, 2018, shareholders are still taxed on ordinary dividends based on their personal income tax bracket, and qualified dividends are taxed as capital gains. Thus, double taxation is still an issue (corporate income tax plus tax on dividends). For the highest tax bracket, double taxation can lead to an effective tax rate as high as 36.8 percent, of which 21 percent would be eliminated in a tax-efficient entity.[32] Thus, in both the high corporate tax era of the 1980s and 2000s and the low corporate tax regime after 2017, it has always been advantageous for corporate timberland owners to have a tax-efficient corporate structure or to sell their timberlands.

As stated, MLPs originally attracted attention within the forest industry because they were a means to separate timberlands to make firms less attractive targets to corporate raiders and to level the playing field in tax treatment while forest products firms still exerted a degree of control over their raw material supply. An added benefit of an MLP spin-off was that it did not trigger a taxable event. This form of timberland sales solved tax issues associated with the income generated from selling timberland.

However, as an investment vehicle, the timberland MLP structure proved to be less attractive to the investment community, particularly institutional investors, because its tax accounting requirements were challenging and cumbersome. Furthermore, rightly or wrongly, some investors came to view the MLP approach as lacking in financial independence because the forest products companies that formed the MLPs often purchased much of the wood from them under preferentially priced supply agreements such as the right of first refusal, and because MLPs continued to manage the land with the same foresters who had managed it prior to their spin-offs. In addition, IP Timberlands and Rayonier Timberlands MLPs basically had a term of fifteen years. In time, both companies bought the minority units of their MLPs that they sold to the public and reintegrated their timberlands with their other assets.

Installment Sales Technique

As noted earlier, the cost basis for industrial timberlands was usually low because of timber depletion and inflation over time. Thus, when a forest products firm sold its timberlands outright rather than doing a spin-off, it often generated a hefty profit, for which a capital gains tax needed to be paid. To reduce this tax liability, some forest products companies used the installment sales technique. This sales method allows partial deferral of any capital gains to future tax years. Installment sales require the buyer of timberlands to make regular payments, or installments, on an annual basis plus interest if installment payments are to be made in subsequent tax years. Often, neither the buyer nor the seller wanted a stream of installment payments, and a bank would step in as an intermediary. Thus, the buyer could pay out a lump of capital to the bank, and the bank would effect what was essentially the fiction of an installment sale.

Because it spreads income over several years, the installment sales technique reduces the tax liability of forest products firms that have sold timberlands. The installment sales technique has thus facilitated the divestment of timberland from industrial owners to institutional investors.

So far, I have discussed tax regulations on treatment of income from timber harvesting and income from timberland sales through spin-offs or outright sales. I have also noted part of the tax implication of the corporate structure of timberland entities—double-taxed C corporations versus pass-through entities. The other part of the tax issue associated with the corporate structure of timberland entities is related mostly to timberland REITs. Timberland REITs issue dividends that are characterized in three ways for the holders of REIT units—return of capital (i.e., depletion), on which no taxes are due; capital gains (gain on timber sales between basis and market), on which lower capital gains tax rates are applied; and ordinary income from logging, manufacturing, or HBU sales, on which ordinary income taxes are assessed.

The Buyers: Realization of Timberland as an Asset and a Historical Chance to Buy Low

As it became clear that forest products companies should sell some of their timberland assets or set up timberland MLPs, they started to search for buyers. Institutional investors were a logical choice because they had large financial capital, they were by nature long-term investors, and land and trees were long-term assets.[33] The development of this new owner class was also helped by several regulatory and market factors.

ERISA of 1974

The Employee Retirement Income Security Act of 1974 (ERISA) was enacted by the US Congress to address failures and irregularities in the administration of some pension plans and to protect the retirement assets of American corporate employees. It created new rules, including asset diversification, that qualified plans must follow to ensure that plan fiduciaries do not misuse plan assets. A fiduciary is a person or organization that acts on behalf of another person or persons with a duty to preserve good faith and trust. Being a fiduciary thus requires undivided loyalty and being bound both legally and ethically to act in the others' best interests. All fifty states have since made their public pension funds conform to ERISA provisions.

ERISA of 1974 defines a fiduciary as anyone who exercises discretionary authority or control over a plan's management or assets, including anyone who provides investment advice to the plan. Those who fail in their fiduciary responsibility may be held responsible for restoring losses to the plan. Specifically, ERISA of 1974 stipulates that "a fiduciary shall . . . by diversifying the investments of the plan so as to minimize the risk of large losses.[34]

In essence, ERISA of 1974 installed the "prudent man rule" in investment fiduciaries such that investments need to be made in a diverse set of securities and assets that are not limited to only stocks and bonds.[35] This enabled investment fiduciaries to invest in real estate. Before ERISA of 1974, most corporate pension funds that had invested mainly in stocks, bonds, government securities, and cash equivalents lost their value in times of high inflation. After 1974, real estate increasingly became an investment asset, and it had attracted more than $20 billion by 1983.[36] Timberlands and farmlands, as a subset of real estate, also started to be on the radar of pension fund managers. It was only in the 1970s that insurance companies such as Travelers and banks such as North Carolina National Bank started to invest directly in timberlands.

CalPERS's Decision to Invest in Timberlands

If ERISA of 1974 was a catalyst for institutional investors to consider real property as an alternative investment option, CalPERS's search for timberland investment advisers in 1984–1986 and eventual decision to invest in timberlands in 1987 signified that timberlands were a viable and promising investment asset for institutions. As noted in chapter 3, CalPERS, the biggest public pension fund in the country, started to invest $150 million in timberlands in 1987. A year later, after a successful performance review, it added another $50 million for a total pledge of $200 million. CalPERS was undoubtedly a market leader in terms of institutional timberland investment in the late 1980s and throughout the 1990s.

Further, and more importantly, because CalPERS was a public institution, it issued a public call for proposals for timberland advisers/managers in August 1985, having previously discussed the matter with potential suitors. All of its investment seminars and interviews with potential candidates for advisers were open to the public. CalPERS's actions were in effect an advertising campaign and an endorsement of timberland investment. Thus, CalPERS helped legitimize institutional timberland investment in the eyes of many institutional fund managers.

It also helped that CalPERS's initial timberland investment paid off in a big way. With skillful guidance and management by Hancock Timber Resource Group (HTRG), CalPERS allocated 60 percent of its initial $150 million in the US Pacific Northwest.[37] The reason, in Rick Smith's words, was that stumpage prices in the Pacific Northwest were lower than those in the South at the time (fig. 2.2)[38] because of the economic recession and timber crisis in the early 1980s. Believing this disparity was only temporary, Rick Smith and his colleagues at HTRG recommended that CalPERS invest heavily in the Pacific Northwest.

A few years later, when the northern spotted owl was put on the nation's threatened species list and afforded full protection in 1990 under the Endangered Species Act of 1973, timber harvests on public lands in the Pacific Northwest were curtailed by 80 percent,[39] and stumpage prices rocketed throughout the country, especially in the Pacific Northwest. CalPERS sold some of its timberland investment around 1996–1997 with a spectacular rate of return of over 30 percent per annum. CalPERS's reentrance into the timberland market in the mid-2000s, however, was much less successful because of high acquisition prices and the prolonged slump in stumpage prices in the US South.[40]

Marketing Timberland as an Investment Asset

Earlier institutional investors such as Travelers and First National Bank of Atlanta (FNBA) invested in timberlands based mostly on their own discovery and experience. However, attracting a large group of institutional investors with no prior experience with investing in timberlands required educating the investors and conducting research on the financial characteristics of timberland investment.

The main challenges were the lack of a track record and the lack of well-accepted valuation methods for timber and timberland assets. Subsequently, the pioneers and early practitioners of TIMOs had to use historical timber prices, timber growth rates, and agricultural land prices as a proxy for forestland prices to do their investment analysis. They had to conduct and present their analysis hypothetically—what the returns would have been had they bought a well-diversified portfolio of timberland across the US South and elsewhere in the 1950s and 1960s. Even though they had reasonable assumptions, used the best data available at the time, and conducted sensitivity analyses, investment in timberland assets was a hard sell.

This contributed to the difficulty of marketing timberlands as an investment asset class. In these early years, there was a lot of timberland and very little capital committed to timberland investment. The situation would reverse a decade later. Since the early 2000s, there has been a lot of financial

capital competing for a limited amount of investment-grade timberland in
the United States.

All pioneering TIMOs did research and marketing work. The two
"Charlies" (Charlie Raper at Travelers and Charley Tarver at FIA) preached
what they called the "gospel of timber." Duncan Campbell thought that he had
to create a market for "a product without market." Robert "Bob" G. Chambers,
who took Charley Tarver's position at FNBA, hired Jim Webb Jr. at Wachovia
(which bought FNBA) in 1987 to strengthen marketing and sales. In 1984,
Rick Smith presented the finding at Hancock that from 1955 through 1983, a
generic timber portfolio consisting of timberland properties in the South and
Northwest would have yielded a total annual rate of return of approximately
14 percent. Smith and others at Hancock marketed timberlands to many corporate and public pension funds.[41] In its analysis, Resource Investment Inc.
(RII) came up with an annual real (inflation-adjusted) rate of return of 6.4
percent for twenty-four years prior to March 31, 1984, and used this result in
its marketing campaign (fig. 3.2).[42]

A few banks—FNBA/Wachovia and US National Bank of Oregon in particular—and an insurance company, Hancock, had actual historical return
data for timberland under their management. Some of them created an index
of their own. Similarly, Wachovia created a Wachovia timberland index, and
FIA had a southern timberland index, both covering only the South after
1977. Hancock, on the other hand, had an index that covered both the South
and Pacific Northwest after 1960.

Based on these indexes or timber prices and forest growth data, these
TIMOs basically told their potential clients that
- timberland investment had competitive rates of return and low risk,
- timberland investment returns were not correlated with stocks and
  bonds, and
- timberland investment matched well with the investment horizon of
  many institutional investors.

The earliest compilation of these systematic marketing efforts toward
timberland investment was done in March 1986 by the Center for Forest
Investment of the Duke University School of Forestry and Environmental
Studies, which sponsored three timberland marketplace workshops around
the country in 1984. These workshops brought timberland buyers, sellers,
investors, and their advisers together. In 1985, it held another workshop that
brought Sir James Goldsmith to Durham, North Carolina. In 1986, it published
proceedings based on the three workshops in 1984.

What also helped these TIMOs market timberland investments was that the real rate of return on pension funds was very low in the twenty years prior to 1980, as shown in figures 3.1 and 3.2. Just as the manager of the trust department at FNBA was searching for an alternative investment vehicle in 1981, many institutional investors were also searching for alternative investment opportunities. In this environment, the market campaign by TIMOs attracted the attention of some institutional investors.

Inefficient Timberland Market: A Historical Chance to Buy Low

When newly emerging TIMOs were trying to secure investors to buy industrial timberlands in the mid-1980s, they also told their potential clients that there was a historical opportunity to buy low. The timber and timberland markets were depressed because some forest products firms were trying to unload a large amount of their timberlands at a large discount. Such a large discount contributed significantly to securing some institutional investors to invest in timberlands. It was also a big reason that returns in the first ten years of the NCREIF Timberland Index were spectacular (table 4.2). Further, the coefficient of variation for timberland was lower than that of the S&P 500, indicating that timberland had a lower level of risk than the S&P 500. If one uses the Sharpe ratio as a valuation parameter, the higher returns and low risk made timberland more than twice as valuable compared to the S&P 500 in the same period.[43]

Indeed, investments in timberland in the 1980s were some of the best because circumstances allowed timberland assets to be available at favorable prices. How depressed was the timberland market in the mid-1980s? Eric Oddleifson of RII believed that "this is the third greatest buying opportunity in the past 50 years. . . . It [industrial timberland] can be gotten at a 40 percent to 50 percent discount, or $400 to $500 an acre."[44]

While the discount associated with the availability of large tracts, many sellers, and few buyers was well understood among timberland market participants, it was subtle and perhaps less known at that time that institutional investors' valuation metrics of these industrial timberlands were different from those of industrial owners for two reasons.

First, industrial timberland owners valued their lands more from a timber production perspective, sometimes neglecting the potential of their lands for HBU. Even though all major forest products companies had real estate development teams that focused on identifying and developing HBU properties within their timberland portfolios and were successful with such

Table 4.2. Annual rates of return for the NCREIF Timberland Index, S&P 500, and US
government bonds, and rates of inflation, 1987–1996

Year	NCREIF Timberland Index (%)[a]	S&P 500 (%)[b]	10-year US govt bonds (%)[c]	Inflation (%)[d]
1987	26.52	5.25	8.39	4.33
1988	30.12	16.61	8.85	4.41
1989	37.35	31.69	8.49	4.64
1990	11.06	-3.10	8.55	6.25
1991	20.26	30.47	7.86	2.98
1992	37.32	7.62	7.01	2.97
1993	22.36	10.08	5.87	2.81
1994	15.44	1.32	7.09	2.60
1995	13.84	37.58	6.57	2.53
1996	10.73	22.96	6.44	3.38
Average	22.50	16.05	7.51	3.69
Standard deviation	9.56	14.03	0.99	1.13
Coefficient of variation	0.42	0.87	0.13	
Sharpe ratio	1.78	0.75	1.41	

a National Council of Real Estate Investment Fiduciaries (NCREIF), accessed December 30, 2012,
https://www.ncreif.org/data-products/timberland/.

b Standard & Poor's (S&P) Index Services, accessed December 30, 2012, https://www.slickcharts.
com/sp500/returns.

c Market yield on US Treasury securities at ten-year constant maturity, Federal Reserve Bank of St.
Louis Economic Research, accessed December 30, 2012, https://fred.stlouisfed.org.

d Bureau of Labor Statistics, US Department of Labor, Washington, DC, accessed December 30,
2012, https://www.bls.gov/data/#prices.

large-scale and complex developments in the 1980s, timber supply was still
their main focus. Institutional investors, on the other hand, focused on total
returns and embraced the opportunity for HBU from the start.

Second, the prevailing timber appraisal method used in the 1980s
was the cost approach, which typically underestimated the value of timber and timberland as compared with the income approach.[45] So, when
RII marketed its timberland fund as seeking an 8 percent real annual rate
of return, some seasoned forest industry consultants were doubtful that it
could be achieved.[46] Nonetheless, different valuation metrics revealed that
inefficiency existed in timberland markets and that a significant arbitrage
opportunity arose between industrial timberland owners and institutional
investors in the 1980s.

A great buying opportunity, indeed. For example, Travelers bought a property in the state of Washington for $6 million in 1987. It sold the property for $26 million in 1993 and generated $2–3 million in timber revenue between these years. The internal rate of return for this investment was easily more than 20 percent.[47]

Because some CalPERS records are publicly available, I was able to find some information on its first land purchase in the South in 1987, which illustrates the undervaluation of timberlands or market inefficiency in timberlands at that time.

It was fee simple (private) land of 90,229 acres with standing timber located between Columbia, South Carolina, and Charlotte, North Carolina, sold to CalPERS by International Paper Company "as part of a corporate strategic plan to redeploy assets." The price tag was $36,470,278, which means that the average price was only $404 per acre. The property was highly productive, and the individual tract sizes were large, averaging four hundred acres each. The property was close to many forest products mills as well as rapidly developing regions in Charlotte. Better yet, Hancock estimated that timber made up over 50 percent of the value of the property as a whole. This property was forecast to exceed Hancock's 8 percent minimum constant dollar hurdle rate for investment in timberlands, with annual cash returns averaging 4.85 percent.[48]

And the great buying opportunity was not limited to industrial timberlands. My former dean, Dr. Emmett F. Thompson, bought a forest property with mature timber near Auburn, Alabama, in the mid-1980s for about $400 per acre. In addition to periodic timber income and recreational activities, his timberland value increased five times in the last thirty years.[49] Two former professors at the University of Georgia bought a small forest property with timber in southern Georgia for $300 per acre in 1990. They paid for their cost of land with one timber thinning a few years later when timber prices were high. In 2013, they sold this property with forests near maturity. For the full twenty-four years of their investments, the internal rate of return was "easily at 20 percent."[50]

All these anecdotal examples point out that there were serious market inefficiencies in timberlands at least in the mid-1980s, and that smart money indeed bought at a historical low. But these inefficiencies had been there for a long time, as other landowners such as Georgia Timberlands had been exploiting and profiting from them for decades.[51]

Professional Management That Adds Value

TIMOs also told their clients that professional management could enhance returns and reduce risk, including market risk, natural risk, and social risk. Professional timberland managers could control and maximize the income stream by harvesting more when timber prices were high or storing timber on the stump where it could continue to grow and increase in value.[52] In other words, they could help institutional investors gain from market inefficiencies in timber and timberlands, and they could reduce risk by diversifying timberland holdings in species, age, and region, or by growing planted forests with genetically improved trees.

Here are two examples of how professional management added value to timberland investors. The aforementioned advice of HTRG to CalPERS to allocate 60 percent of its initial capital to timberland investment in the Pacific Northwest, against CalPERS's initial asset allocation plan of 30–40 percent, was based on the intertemporal inefficiency that existed in different regions.[53] Today, TIMOs have reached out to many countries on all continents where trees can grow. This is a magnified example of exploring regional differences at a global scale. Of course, global timberland investment has added challenges. Chapter 7 is devoted to international timberland investment by institutions.

The other example, known to many foresters, is that the value of timber-growing lands increases faster in certain periods of forest growth and can be a source of higher investment returns, as illustrated in figure 2.3. The incremental growth in timber value during that period, from year t_1 to year t_2 in figure 2.3, is substantially higher than in other periods. Typically, this period could be ten to fifteen years for southern yellow pine forests and ten to twenty-five years for Douglas fir forests.

In a perfect market, no growth period should generate better returns than any other growth period, assuming the purchase price equals the discounted value of future revenues. This implies that purchasers apply a higher discount rate to properties in the rapid-growth period. Yet timber and timberland markets are imperfect, as illustrated in chapter 2, because roughly the same discount rate is used by timberland investors across the United States irrespective of timber age and species composition. A related point is the argument that faster-growing trees provide better returns than slower-growing ones. This is not necessarily true, and, as demonstrated in many examples throughout this book, it all depends on acquisition price.

RII tailored its first few funds to timberlands that were primarily in this high-value growth period and secured an annual rate of return of more than

10 percent for these funds. Hancock adopted a similar strategy in the 1980s. Today, some timberland portfolio managers still craft and implement strategies that exploit this inefficiency, such as buying young forests at a heavy discount, holding them for ten to fifteen years, and then selling them at a premium when trees are nearly mature. Granted, this inefficiency has now been exploited by so many TIMOs that it may be hard to find a way to implement it at scale.

Finally, there were cases of corporate executives' self-interest being placed over shareholders' interest in the sales of some industrial timberlands as far as the shareholder value school of corporate governance is concerned. In December 1991, a TIMO got a phone call from a major forest products company that wanted to sell $50 million of timberlands in twelve to fourteen days, before the end of the year. Apparently, that company was short of its earning target for the year. The quickest way for the company to generate income in this situation was to sell some of its timberlands, because the accounting basis for its timberlands was below the fair market value. The TIMO's reaction was that it would be difficult because it was a holiday season and it had to seek institutional investors who were willing to commit that amount of money (which was large in 1991) in a short period. Eventually, the company kept sweetening the deal by adding more acreage—and thus lowering the average price per acre—until an investor finally jumped on board and bought the timberlands before year's end.[54]

In addition to all these spatial and temporal inefficiencies in timber and timberland markets, including imperfect information among others, large tracts of timberland tended to be sold at a discount at least in the 1980s and 1990s, giving the much more limited universe of qualified purchasers and the possible transaction costs to subdivide them. Thus, there might be a diseconomy of tract size in timberland sales. Nowadays some people think this "wholesale discount" has disappeared and there might even be a "wholesale premium" because of competition for large assets. Also, TIMOs were very happy to do conservation sales and easements, whereas until lately, integrated forest products firms were not. In all events, the reliance by institutional investors on professional management also helped them find investment opportunities and enhance their investment returns.

An Unprecedented Arbitrage Opportunity for Industrial Timberlands

When institutional investors entered the timberland markets in the mid-1980s, they had a different valuation metric and valued timberlands more than industrial owners. Their valuation was enhanced by several facts: they would pay only a single dividend tax, they could manage and harvest timber with flexibility and pursue the most profitable markets without being tied to any specific mills, and they could more aggressively explore the possibility for HBU and conservation sales. All things considered, they perceived that timberlands could not only generate competitive returns but also offer diversification benefits to their investment portfolios.

On the supply side, forest products firms were under duress because of the recession in the early 1980s and the prolonged crisis in the forestry sector that followed. In order to fend off potential hostile takeovers, nearly all forest products firms either monetized some of their timberland assets through outright sales or timberland MLPs, or had to treat their timberlands as an independent profit center. Their motivation to sell timberlands was further amplified because forest products manufacturing could be profitable without owning timberlands, the costs of holding these lands were too high, their shareholders paid double taxes, and they lost the capital gains tax benefits for timber income in 1986. Most importantly, forest products firms had, based on GAAP, depleted the timber resources on their timberlands over many years and thus had a low cost-basis for their timberlands and could quickly profit from divesting these timberland assets.

Thus, a historical arbitrage opportunity arose, and the transition from industrial to institutional timberland ownership began in the mid-1980s and escalated in the following two and a half decades. The poor timber market conditions in the mid-1980s, a flood of timberlands on markets, timber and timberland market efficiencies, the marketing efforts of TIMOs, and the advice of Wall Street all helped expedite this process.

CHAPTER 5
The Growing TIMO Trees

In chapter 3 we saw the multiple origins, pioneers, and early practitioners of the TIMO business up to 1987. This chapter focuses on the contemporary development of TIMOs and their record in forest management and conservation in the last thirty-plus years. To grow their business, several TIMOs engaged academicians to study the economic and financial characteristics of timberland investment. As more institutional investors recognized timberlands as a viable investment vehicle and began to search for and buy timberlands, they needed more TIMOs to manage their assets. Consequently, TIMOs grew quickly in number and in total assets under management. A few pioneering TIMOs split into separate companies. New TIMOs emerged. Several TIMOs arose and disappeared because of mergers and acquisitions, changes in key personnel, and competition. On the other hand, the success of the Hancock Timber Resource Group (HTRG) as a business alliance perhaps contributed to its breakup in 1997. In this period, we also witnessed the largest private endowment in the United States—Harvard University Endowment—entering timberland markets in a big way, profiting greatly, and then exiting timberland investment with a write-down.

As virtually all industrial timberland changed ownership with institutions via their TIMO managers controlling the majority, three questions immediately arose. First, would these new owners continue the intensive management practiced by the prior industrial ownership and therefore secure long-term timber supply for industrial activities supporting rural communities? Second, would the new owners manage their forests in an environmentally responsible way? Finally, would they support forestry research and development (R&D) as the previous owners had? Therefore, this chapter also discusses forest management and conservation under new institutional ownership, a great concern to the forestry and environmental community, as well as timber supply agreements between institutional owners and forest products companies. The third question, however, is much harder to answer, as there is no current systematic study on the R&D expenditure of institutional owners. Nonetheless, I try to address this question with anecdotal evidence. The first part of this chapter is by and large descriptive, but no less interesting from a historical, economic, or industrial organizational perspective.

Academic Studies Supporting Timberland Investment

The earliest publication by academicians on timberland investment using a portfolio approach was probably by W. L. Mills and William L. Hoover in 1982, although their article sought to explain why many nonindustrial private land-owners owned forests despite traditional analyses yielding low net present value or internal rates of return for forestland as a stand-alone asset. They found that investment in a timberland asset (a hardwood forest in the northern United States) provided a diversification benefit for landowners. In 1988, Clair H. Redmond and Frederick W. Cubbage found that, based on data on historical regional stumpage prices as a proxy for stock appreciation and the net regional timber growth rates as the "dividend," for any level of risk assumed, timberland exhibited returns slightly better than stock market returns.[1] Further, timberland returns generally moved countercyclically to stock market returns.

In the mid-1980s, some newly established TIMOs felt that they needed independent academic studies on timberland investment. Consequently, they started to hire academicians to write refereed articles. Wachovia Timberland Investment Management, a TIMO owned by the First Wachovia Bank and Trust, along with the US Forest Service and a few forest products firms, provided financial support to some of these studies.

The first article published from this effort was written in 1989 by Robert Conroy of the University of Virginia and Mike Miles of the University of North Carolina.[2] They were hired because of their experience in setting up the real estate index for the National Council of Real Estate Investment Fiduciaries (NCREIF). Then, a few more articles were published from 1989 to 1991. The authors were F. Christian Zinkhan, who started to research the subject as a graduate student at Duke University; Courtland Washburn, a graduate student at Yale University; and Clark Binkley, then a professor at Yale University.

These studies confirmed the attributes of timberland investment claimed by the pioneers of TIMOs:

- timberland investment generates returns comparable to the returns of broad stock markets;
- timberland investment has low systematic or market risk, which means that timberland investment returns are not correlated with stock returns and thus provide an opportunity for investors to diversify their portfolio and improve its efficiency;
- historically, timberland investment had a high and significantly abnormal rate of return, possibly related to market inefficiency, wholesale discount, and illiquidity of timberland assets;

- timberland investment returns have low volatility in terms of standard deviation, which is used as a measure of risk; and
- timberland investment hedges inflation.

In the next few paragraphs I summarize the method used in these earlier studies and explain terms such as market risk (beta) and abnormal rates of return. Those who are familiar with the method or do not want to get into technical details can skip these paragraphs.

* * *

Most of the earlier studies on timberland investment used the Capital Asset Pricing Model (CAPM). Developed by William F. Sharpe and John Lintner, CAPM states that the required or expected return on an investment should be equal to the rate earned on a riskless investment plus a premium for the assumption of market risk:[3]

$$R_i = R_f + \beta_i (R_m - R_f) \tag{4.1}$$

where

R_i = the required or expected rate of return on investment i;

R_f = the risk-free rate of return (measured by the short-term yield on US government Treasury bills);

β_i = investment i's risk premium, commonly known as beta, which is equal to the covariance of R_i and R_m divided by the variance of R_m; and

R_m = the market's expected rate of return (often measured by a broad stock market index).

Because expected returns cannot be observed, the model must be estimated from ex post data. Michael J. Jenson proves that CAPM is consistent with the regression equation or excess return form:[4]

$$R_i - R_f = \alpha_i + \beta_i (R_m - R_f) + \mu_i \tag{4.2}$$

The intercept α_i (alpha) in CAPM signifies the return of investment i due to factors other than the overall market. A positive alpha indicates that the investment has an expected return greater than the market requires in the risk class (as measured by beta) and thus demonstrates a superior risk-adjusted return or abnormal return. Beta (β_i) indicates the investment's market risk or systematic risk. If the beta value is greater (smaller) than one, the investment moves more (less) than a corresponding move in the whole market. Thus, such an investment is said to be more (less) risky than the market.

Because there was no systematic return data on timberland investment in the country in the 1980s, researchers used various ways to compute ex post returns for timberland investment, including stumpage prices only, stumpage prices plus timber growth, and stumpage prices plus timber growth and land appreciation using agricultural land appreciation as an approximation.[5] Furthermore, as noted earlier, the FIA southern timberland index, Wachovia timberland index, and Hancock timberland index were created by these TIMOs based on the performance of properties or funds they managed.

In 1989, HTRG sponsored a discussion paper by Pension Realty Advisors Inc. of San Francisco and produced a second edition of it in 1991. Titled "Timberland: An Industry, Investment, and Business Overview," this paper reviewed the history and attributes of timberland investment and provided some educational perspectives on the unique investment and operational elements of timberland assets. One conclusion of this discussion paper is that the investment scale of timberland contributes to its illiquidity, because "timberland investment markets operate on the basis of imperfect information, long-term investment horizons, and the ability to conduct transactions at opportune times."[6] In the following decades, various TIMOs used this paper and other gray literature as well as their own brochures to market timberland investment.[7]

In 1992, F. Christian Zinkhan, William R. Sizemore, George H. Mason, and Thomas J. Ebner published a book titled *Timberland Investments: A Portfolio Approach*, which synthesized all threads of timberland investment practices and research up to 1990. It was a culmination of timberland investment knowledge at the time. This book has since become a must-read on timberland investment and helped attract institutional investors to invest in timberlands.

Around 1990–1991, CalPERS advised HTRG to develop an independent and industry-wide measure of timberland returns based on the actual performance of managed timberland properties. Hancock then teamed up with FIA and PruTimber (which had just become a TIMO in 1990), the Frank Russell Company, and NCREIF and developed the NCREIF Timberland Index in early 1992. After more than two years of preparatory and review efforts, NCREIF commenced publication of the NCREIF Timberland Index in late 1994. With data back to the fourth quarter of 1986, this index has since become the benchmark and leading performance indicator for the TIMO business. Other pioneering TIMOs at that time, namely Travelers, Wachovia, and RII, did not initially participate in or contribute data to this index for competitive and other reasons, but most of them joined several years later.

Canal Industries and Wagner Forest Management were not included, probably because they were primarily wood dealers and forestry consultants
and their fiduciary timberland investments were rather small at that time.

The NCREIF Timberland Index is based on various timberland properties whose number, total value, income, and average property value per acre
are reported. The total returns of the NCREIF Timberland Index are decomposed into two elements. The first, referred to as EBITDDA (earnings before
income tax, depreciation, depletion, and amortization), is an income return
or cash dividend, reflecting the current net operating revenues associated
with timber harvest and sales of various nontimber products. The second is
an appreciation return, reflecting the change in the value of the timberland
property, including the bare land and timber inventory. Only total returns
are used in most studies that use this index.

The Growing TIMO Trees

As noted in the previous chapter, at the beginning of 1987, there were seven
TIMOs—Travelers, FIA, HTRG, Wachovia Bank, Resource Investments Inc.
(RII), Canal Industries, and Wagner Forest Management. At that time, Canal
Industries was mostly a wood dealer, and Wagner Forest Management was
primarily a forestry consultant and forest property manager. Similarly,
Forest Resource Consultants, a consulting firm in Macon, Georgia, managed
the forest property for British Coal Company. Another company, Fountain
Forestry, a subsidiary of a UK firm, also had some investments in US timber-
lands for some private UK citizens and institutions.

In 1989, Prudential Insurance Company, which at that time was the
largest insurance company in the country, decided to have a TIMO. Like
Travelers and Hancock, Prudential had been in the agricultural and forestry
loan business for a long time. Prudential wanted to copy Hancock's timber
business and recruited John Lord, senior portfolio manager, and Douglas
Charles, forestry investment officer, from Hancock in January 1990 to set up
a TIMO subsidiary called PruTimber. PruTimber grew over the next sixteen
years, and by 2005 it had $1 billion in timberland assets under management.
In 2003, it even considered buying HTRG. Ultimately, it went the other way
around: Hancock bought PruTimber in 2005.[8]

In 1993, The Lyme Timber Company started to offer limited partnership units in a commingled fund to institutional investors and thus officially
became a TIMO. The Lyme Timber Company has since launched five commingled funds and managed property from New England to Florida. It has

invested mostly in properties that have conservation value and generate a portion of their returns through sales of conservation easements.

Two more TIMOs emerged in 1995. One was The Forestland Group (TFG), founded by F. Christian Zinkhan. Zinkhan and four other partners had tried to start a TIMO three years earlier, but when that effort failed, Zinkhan went alone and succeeded in establishing TFG. Based in Chapel Hill, North Carolina, TFG focuses on investment in hardwood forests. By 2020, it had become one of the largest TIMOs in the United States and had investments in Belize, Canada, Costa Rica, and Panama.

The other was Timberland Investment Services (TIS) LLC, which was founded by Robert Chambers and two partners. Chambers was hired to replace Charley Tarver at First National Bank of Atlanta in 1984. After Wachovia bought First National Bank of Atlanta, Chambers continued to work as the head of Wachovia Bank Timberland Investment Management Group until 1994. TIS later changed its name to Timbervest for branding and marketing reasons.

Timbervest was based in Woodstock, Georgia, from 1995 until April 2003 when Chambers sold his share and left the company. The new owners and managers moved the office to Buckhead in Atlanta, Georgia. In 2015, Timbervest was ordered to pay $585,000 and its principals were barred from the investment advisory business for five years for two separate land deals in 2006 and 2007 in which these principals profited from undisclosed real estate commissions.[9] Although Timbervest denied the charge and vowed to continue to fight the case in federal court, it was handcuffed by the five-year ban. In May 2017, Timbervest was sold to Atlanta-based Domain Timber Advisors LLC.

As noted earlier, RII was sold to Union Bank of Switzerland in 1995 with $0.8 billion in assets under management. In late 1996, Eric Oddleifson responded to an invitation from Jeremy Grantham to establish a timber venture with Grantham, Mayo, Van Otterloo & Co. (GMO). In forming the partnership, Oddleifson recruited his joint venture partner Eva Greger, who had worked closely with him since 1985, with the understanding that she would succeed him in building the new TIMO and lead the management team after his retirement. So, GMO Renewable Resources was created in 1997 and became a significant TIMO in the United States and elsewhere. In 2017, following Jeremy Grantham's retirement, GMO sold its renewable resources capability to The Rohatyn Group (TRG), which continues to offer timber and agricultural investment strategies.

Also in 1997, Pope Resources MLP formed a wholly owned subsidiary, Olympic Resource Management LLC, as a TIMO. However, it was not until 2003 that Olympic Resource Management launched its first timberland private equity fund and started to acquire and manage properties on behalf of third-party investors. Pope Resources coinvested alongside the institutional investors in all investments. Pope Resources MLP as well as its subsidiaries was purchased by Rayonier in January 2020.

* * *

One of the most significant events in the TIMO business in 1997 was the breakup of the business alliance in HTRG. HTRG was created to attract institutional timberland investment business, especially in anticipation of a call for proposals for timberland investment advisers from CalPERS in 1985. In the following decade, it succeeded in getting timberland investment business from CalPERS and many other institutions. With more than $2 billion in assets under management, it was the largest TIMO in the world in 1997.

In its proposed timber investment program for CalPERS's RFP No. 85-31, HTRG stated that it was a consortium of proven, independent performers having fiduciary and timber investment expertise and a cohesive, focused

Figure 5.1. Organizational chart of Hancock Timber Resource Group delineating management responsibilities

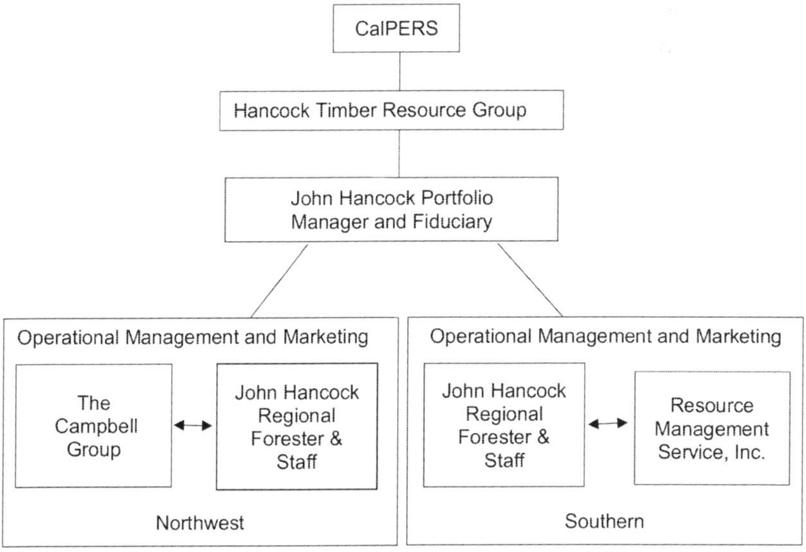

Source: Hancock Timber Resource Group (1986).

organization consisting of the John Hancock Insurance Company, Resource
Management Service Inc. (RMS), and The Campbell Group. Operationally,
Hancock was responsible for all timber investment management recommendations to CalPERS, while The Campbell Group and RMS would work
closely with Hancock's regional staff to perform the acquisition analysis and
operational management as directed by Hancock. Hancock was the portfolio
manager and fiduciary, and the other two firms were the regional arms (fig.
5.1).[10] Hancock and each of the other two firms signed simple and separate agreements to formally establish HTRG in the summer of 1985. In the
agreement, either firm could dissolve the partnership if performance was
not satisfactory. In addition, this was an exclusive relationship, meaning that
the other two regional firms could not elicit institutional investors in asset
acquisition and management. The management and incentive fees would be
divided between Hancock and the other two firms.

This organizational structure added value in the eyes of institutional
investors in the early days of institutional timberland investment. HTRG
also helped legitimize the TIMO business model. In the first few years, this
arrangement worked well; all three parties worked hard together to make
their fledging business alliance succeed. Gradually, different views on corporate strategy for HTRG as a whole and for individual companies got in the
way, as did issues regarding further delineation of management responsibilities and fees, personality, greed, control, and egos.

One way for each company in the consortium to earn more money was
to increase the fees that HTRG charged to its clients. HTRG decided not to do
so for competitive reasons. Then there was discussion about how to define
each partner's role. Although the former executives of all three firms did
comment on my hypothesis, I sensed that the two regional partners were
somewhat handcuffed because they could not elicit investments from institutional investors and could not grow their acquisition and asset management businesses, let alone become an independent, new TIMO if they ever
aspired to do so. Because nearly 100 percent of The Campbell Group's business and some 90 percent of RMS's business was tied to HTRG, these two
firms might have wanted to diversify their business instead of putting all
their eggs in one basket. Hancock, on the other hand, with an increase in
the number of skilled people it had with a forest operations background,
might have wanted to become vertically integrated in order to receive all
management fees. At a minimum, Hancock could have objected to some
management activities, thereby making more decisions or suggestions on

acquisition and management than before, potentially overstepping into the other partners' territory in asset acquisition and management and threatening the very existence of these regional partners. The delicate balance among the three partners started to crack.

The leaders of the three partners held several discussions between 1992 and 1997. It thus seemed that each company was thinking about its own corporate development strategy and that there was friction among them. This was followed by a mutual recognition that the alliance might have served its purpose and that it might be better to change or dissolve it.

The final straw that broke up HTRG as a consortium was the demand from Hancock that the other two firms became vendors of Hancock, to which The Campbell Group did not agree. So in 1997, the business partnership between Hancock and The Campbell Group was dissolved. The name of the alliance, Hancock Timber Resource Group, remained with Hancock.

A lawsuit in a federal court ensued between The Campbell Group and Hancock over the significant value of incentive fees that had built up mostly in the Pacific Northwest.[11] These properties were managed by The Campbell Group. The Campbell Group claimed that Hancock had not properly calculated the incentive fees and that it was entitled to a greater share (than RMS). Hancock counterclaimed. The case was eventually settled out of court in 1998, and each party got part of what it wanted. Because of a nondisclosure clause in the settlement, all parties remain silent about the case.

So, in 1997, The Campbell Group became an independent TIMO again, while RMS stayed with Hancock until December 2003, when it became a TIMO on its own as a limited liability company (LLC). Nonetheless, Hancock and RMS redefined their relationship after 1997: Hancock started to expand more into timberland acquisition and management, and RMS was allowed to contact and manage assets for tax-paying institutions and private investors, but not for tax-exempt institutions.

Partly because of its experience in the Pacific Northwest where there was a boom in timberland investment, it did not take long for The Campbell Group to regroup and even become the largest TIMO in terms of assets under management at one point in the early 2000s. The breakup of the HTRG business alliance also affected Hancock, as Rick Smith, the architecture and engineer of HTRG, was let go and then founded a new TIMO called Forest Systems LLC a year later, and Clark S. Binkley, professor and dean of forestry at the University of British Columbia, was hired as the chief investment officer. Hancock expanded its on-the-ground operations in asset management and has been the most vertically integrated TIMO in the United States ever since. This breakup also affected some of HTRG's clients. For example, CalPERS decided to issue a request for proposals to put its timber portfolio with HTRG out for bid in 1998, although Hancock eventually won the majority of it back.

In retrospect, the breakup of the HTRG business alliance was perhaps a natural evolution of an industrial organization. The HTRG business alliance added value at the beginning of the TIMO business when institutional timberland investment was largely unknown and no single TIMO had a track record of managing the $150–200 million investment that CalPERS intended to allocate. As the timberland asset class developed and business increased, all three firms wanted to grow their businesses and control their own future, and thus they could have overstepped the thinly defined boundary among them in terms of advising, client recruitment, portfolio management, acquisition, asset management, and daily contacts and interactions with institutional investors. As the TIMO business expanded and become increasingly competitive, the alliance provided little additional value. These disputes over corporate strategies, management responsibilities, and fees as well as personality and desire for control ultimately destroyed the delicate power balance between Hancock and The Campbell Group in 1997, and between Hancock and RMS in 2003. Wagner Forest Management, the latest partner added to HTRG in 1990, also returned to sole operations in 2004.

The decision to dissolve HTRG as a business alliance, even though it might have been painful to some employees in these firms at the time, was the right thing to do for the development of the TIMO industry in the United States and elsewhere. All three firms have done well so far, which is the best empirical evidence and an ultimate testimony to justify the breakup of the business alliance. Without doubt, HTRG as a business alliance and the subsequent three independent firms contributed greatly to the evolution and diversity of TIMO organizations in the United States and elsewhere.

* * *

The year 1997 also marked the entrance to timberland investment by the largest private endowment in the United States—the Harvard (University) Endowment. The Harvard Endowment, which was managed by the Harvard Management Company (HMC), had been courted by several TIMOs for years prior to 1997. Previously, HMC had not been involved in the timberland business for two reasons. First, its hurdle rate was very high, and timberland investment in normal circumstances could not reach or surpass its hurdle rate. Second, HMC managed most of its assets in house and did not want to hire a TIMO if it was to get into the timberland business.

L. Richard Doelling, a principal at Forest Investment Associates Inc. (FIA), was among those who had communicated with HMC regarding possible investment in timberlands. In December 1996, Doelling got a call from Jane Mendillo, who headed up alternative investments for HMC. Jane Mendillo asked Doelling whether he could come and speak with her and her colleagues at HMC about timber after the Christmas holidays. Clearly, HMC was thinking about timberland investment, or it might have lowered its hurdle rate, making timberland investment a possibility at the end of 1996.

In January 1997 Doelling went to Boston and met with Jane Mendillo and Jack Meyer, then president of HMC. Jane Mendillo gave Doelling a heads-up that Jack Meyer did not like presentations but preferred to ask questions. Doelling told me later: "He asked, by far, the most astute questions I had ever heard from a prospective investor and two hours later he was done. On my way over to Albany, New York, to meet with one of our clients, Jane called to tell me that HMC wanted to invest $50 million with FIA in a separate account."[12]

Doelling and Mendillo thus began to work on an investment management agreement and explore various investment opportunities. In 1998, they reviewed a hardwood property in northwestern Pennsylvania that International Paper was selling. Richard and his colleagues at FIA liked it and were convinced that it was going to be a very good-performing asset if they could buy it at the right price. It was a very large property and was priced at about $200 million. Doelling spoke with Mendillo of HMC and told her that FIA might like to put more than the normal 10–15 percent of HMC's commitment into this single deal if FIA was successful in purchasing the property. In the end, HMC said that if FIA liked the property that much, it would like to buy the whole property. That ended up being exactly what happened, and today it remains one of the top-performing assets that FIA has ever invested in for its clients.

The Harvard Endowment increased its investment in timberland to $1–2 billion, or 10 percent of its capital, in the early 2000s in the United States and other countries. Because of its sheer size, HMC initially used several TIMOs—FIA, The Campbell Group, and GMO, among others—to handle its timberland investment. Gradually it developed in-house expertise and managed most of its timberland investment internally. As we will see in chapter 7, the Central Island forestry investment in New Zealand was another "home run" by Jane Mendillo and Jack Meyer. It led to the hiring of Andy Wiltshire at HMC and the incredible growth of their in-house capability. In the last few years, HMC decided to ultimately unwind its timberland (and agriculture) portfolio with the write-down of tens of millions of dollars of unrealized gains after paying departed managers tens of millions of dollars based in part on those valuations.

* * *

In 1998, Jeff Nuss founded GreenWood Resources, which started as a property manager for GMO and then became a TIMO in 2007, focusing on plantation forestry investment. In August 2012, GreenWood Resources was acquired by TIAA-CREF.

In 1999, Molpus Woodlands Group started to attract institutional investment and became a TIMO even though its predecessor, Molpus Timberlands Management, had existed since 1996.

In 2002, Forest Capital Partners was established as a TIMO by former Hancock employees Scott Jones and Matt Donegan and was sold, along with property under its management, to HTRG in 2012. The principals of Forest Capital Partners had been involved in the Timberland Growth Company, set up by Potlatch to explore publicly traded REITs in 1997, which will be covered in next chapter.

In 2003, Peter Mertz along with his partners cofounded Global Forest Partners LP (GFP) to acquire the timberland investment capability from UBS RII, which Mertz had led as chief investment officer and managing director since 1997 and which had grown 50 percent in terms of total assets under management. Now operating as an independent, employee-owned investment advisory firm, GFP built on its pioneering investment strategies, focusing on international timberland investment and nearly tripling in size under Mertz's leadership over the following fifteen years.

Here are a few more TIMOs established after 2002:

- Timberland Investment Resources, created in 2003 by Mark Seaman, who had succeeded Bob Chambers at Wachovia Bank after 1995 and departed when Wachovia decided to sell its timberland business to Regions Bank (RMK).
- Regions Morgan Keegan (RMK) Timberland Group, which was created in 2006 by Regions Bank after it purchased the timberland business

from Wachovia Bank. RMK was sold to BTG Pactual (a Brazilian invest-ment bank) in 2013 and now operates as BTG Pactual Timberland Investment Group.

- Resource Management Service became a TIMO in January 2004, as noted earlier.
- Conservation Forestry became a TIMO in 2004 (and changed to Conservation Resource Partners in 2017).
- Pinnacle Forest Investments became an independent TIMO after PruTimber was purchased by Hancock in 2005.
- Jamestown Invest, a real estate investment and management company, started a TIMO business in 2009.
- The Rohatyn Group bought GMO Renewable Resources in 2017, as noted earlier.
- Silver Creek Capital Management became the sole owner and portfo-lio manager of a TIMO business in 2017. Silver Creek was the man-aging partner of the joint venture called Twin Creeks Timber LLC with Plum Creek (a REIT) in 2015. When Plum Creek's successor—Weyerhaeuser—decided to sell the original shares owned by Plum Creek, Silver Creek Capital found a private investor to buy them in 2017.

In addition, TIMOs based in other countries have been created, and some traditional forest management companies such as American Forest Management and F&W Forestry have added some "TIMO-like" businesses—mostly in property management. Chapter 7 covers global timberland invest-ment by both US-based and non-US-based TIMOs.

Figure 5.2 shows the TIMO family trees between 1980 and 2020. Table 5.1 provides a list of the top twenty-five TIMOs in the United States and else-where in 2020. It is evident that large TIMOs are based mostly in the United States.

Alignment of Investors' Interests with TIMO Management

So, in a few decades of institutional timberland investment, we have wit-nessed a proliferation of diverse organizational structure in TIMOs. As investment advisers, all TIMOs have a business model based largely on an annuity of annual asset management fees, which have some combination of three components: fiduciary and portfolio management, asset acquisition and management, and on-the-ground operations such as fire protection, site preparation, reforestation, thinning, timber sales, and HBU development.

Figure 5.2. TIMO family trees

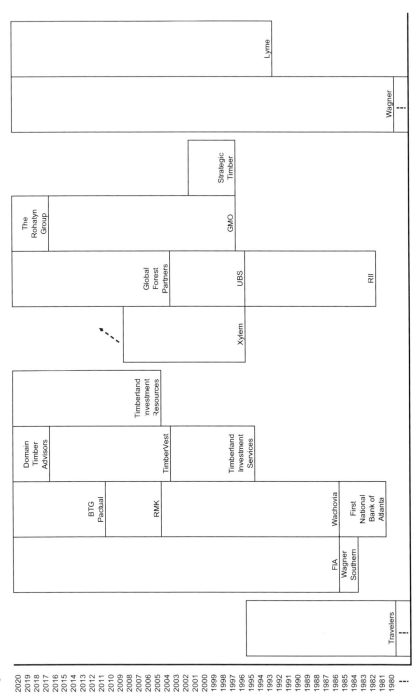

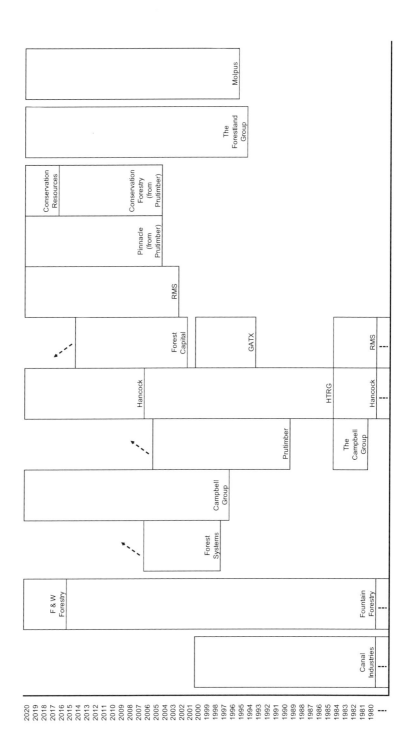

Table 5.1. The top twenty-five TIMOs in the world by assets under management in billions of US$ and millions of acres as of Dec. 31, 2020

Rank	Name	Value	Acres	Headquarters	Geography	Source
1	Hancock Timber Resource Group	10.3	5.47	US	US, Canada, Australia, New Zealand, Chile	htrg.com
2	New Forests	5.0	2.35	Australia	Australia, New Zealand, Malaysia, Indonesia, US, Laos	newforests.com.au
3	Forest Investment Associates	4.6	2.28	US	US, Brazil, Chile	forestinvest.com
4	Resource Management Service	4.2	2.3	US	US, Brazil	resourcemgt.com
5	BTG Pactual Timberland Investment Group	3.5	2.6	Brazil	US, Brazil, Uruguay, South Africa, Hungary	timberlandinvestmentgroup.com
6	Campbell Global[1]	NA	1.82	US	US, Australia, New Zealand, Chile	campbellglobal.com
7	Global Forest Partners	2.7	1.36	US	Brazil, Uruguay, Chile, Guatemala, Colombia, Australia, New Zealand, Cambodia	gfplp.com
8	Stafford Capital Partners[2]	2.6	NA	UK	Various countries	staffordcp.com
9	Gresham House	2.45	0.334	UK	UK, Ireland	greshamhouse.com
10	Société Forestière de la Caisse des Dépôts	2.2	0.74	France	France	forestiere-cdc.fr
11	Molpus Woodland Group	2.14	1.9	US	US	www.molpus.com
12	Wagner Forest Management[1]	NA	2.3	US	US, Canada	wagnerforest.com
13	The Rohatyn Group (ex GMO)	1.93	1.33	US	US, Uruguay, Australia, Chile, Brazil, New Zealand, Costa Rica, Panama	rohatyngroup.com
14	Timberland Investment Resources	1.72	0.836	US	US	tirllc.com
15	The Forestland Group	1.7	2.54	US	US, Panama, Belize, Costa Rica, Canada	forestlandgroup.com
16	GreenWood Resources	1.54	0.57	US	US, Chile, Brazil, Colombia, Poland	greenwoodresources.com
17	Lyme Timber	1.5	1.2	US	US	lymetimber.com
18	Dasos Capital	1.1	0.754	Finland	Finland, Estonia, Latvia, Lithuania, Ireland, Portugal, Poland, Romania, Malaysia	dasos.fi
19	Silver Creek Capital Management	1.03	0.526	US	US	silvercreekcapital.com
20	Teak Resource Company (ex Floresteca)	0.9	0.435	Brazil	Brazil	teakrc.com
21	Brookfield Timberlands Management	0.8	0.675	US	Canada, Brazil, US	bam.brookfield.com
22	Conservation Resource Partners	0.74	0.521	US	US	conservationresources.net
23	Olympic Resource Management (Rayonier)	0.55	0.141	US	US	rayonier.com
24	Domain Timber Advisors (ex Timbervest)	0.64	0.22	US	US	domaincapitalgroup.com
25	Lone Rock Timber	0.51	0.13	US	US	lonerockresources.com

1 NA = not available. Neither Campbell Global nor Wagner Forest Management provided its respective AUM. Thus, their AUM and rank are based on the author's estimates.
2 Since Stafford is in mostly secondary markets, its AUM may double-count those of other TIMOs.

To date, no TIMO has the full in-house operations necessary for ordinary on-the-ground work, and all rely on local consulting foresters and independent contractors to do at least some of this work. As for asset acquisition, some TIMOs charge an acquisition fee, much like the real estate fees charged by realtors, especially at the beginning of the TIMO business. For asset management, the fees may be based on acreage, or the total value of the asset under management, or the total cost of management activities in a given year. As stated earlier, many banks, forestry consultants, and wood dealers had managed timberland assets for private landowners long before TIMOs became a business and have continued this practice. Some forestry consultants have standing on-call agreements with TIMOs and thus essentially act as an extension of the office and personnel of TIMOs.

What distinguishes TIMOs from the forest asset management performed by forestry consultants is their fiduciary and portfolio management responsibilities. All TIMOs must be qualified financial investment advisers. Today, the best of them are SEC-registered investment advisers. Thus, we see some TIMOs developed from top (fiduciary and portfolio management) to bottom (asset management), such as Hancock and FIA, and vice versa, such as Wagner Forest Management.

Institutional investors understand that investments through TIMOs are a long-term commitment and want to secure total returns. TIMOs therefore are not under pressure to pay dividends each quarter. In this aspect, the interests of investors and TIMOs are well aligned. On the other hand, because TIMOs' business model is based on an annuity or annual management fees, it makes sense for TIMOs to raise money, have a large asset base, and hold on to assets a bit longer than they would otherwise. It is at this juncture that the interests of institutional owners and the interests of some TIMOs may diverge, especially in commingled funds.

This mismatch of interests happens in two areas. In the first, if a TIMO has raised money, its investors typically like to deploy the capital within a certain period. If not, they might withdraw their commitments or find another TIMO to invest with. This could invite aggressive or inflated bidding on certain assets by TIMOs that want to deploy an investment quickly. In the second, TIMOs would like to manage assets as long as possible, especially when they cannot sell the property at the appraised or desired value.[13] When an institutional investor puts money in a commingled account with multiple investors, it may be difficult to exit.

Because of poor timber market conditions after the 2007–2009 financial crisis, the managers of some near-expiring timberland funds have offered their investors extensions and usually got them for good reasons. Nonetheless, some managers are doing acrobatics to find ways to hold on to assets. Here lies a big difference between investments in commercial real estate and timberlands. Commercial real estate valuations are based on rents or cash flows. A commercial property is either rented out or its valuation has to come down. In timberlands, current income or cash flow is less of a proxy for the actual value. Low levels of current cash income could imply low levels of income in the future, or they might reflect the timberland manager's choice to forgo current income in favor of higher cash flow later. It is thus difficult for investors to judge the performance of their timberland managers, because a low current yield does not always imply a low total return, and vice versa.[14]

Forest Management under TIMOs

Forest management matters to policy makers and the public as well as to investors and landowners. Many observers wonder whether institutional investors operating under TIMOs' advice manage their forests differently from industrial and nonindustrial owners and whether their management fosters forest sustainability.

To answer these questions, I searched for empirical evidence of forest management behaviors among various owners. If these institutional timberland owners do not overharvest and do reforest after timber harvest, they pass the simplistic test of forest sustainability. A more sophisticated question addresses whether they conserve forest resources. An even more challenging question—whether they are more likely to convert their forests to higher and better uses—probably cannot be answered, as I have not found any empirical study on this subject.

Studies on forest management behaviors among different groups of landowners are done mostly in the US South, the largest timber-producing region, accounting for some 58 percent of timber supply in the country.[15] Researchers have found that individuals (often classified as nonindustrial forestland owners) owned some 54 percent of all timberlands in eleven southern states in the 2005–2011 US Forest Service forest inventory measurement cycle (table 5.2). TIMOs managed some 8.8 million acres, or 5.1 percent, which was five times more than the 1.39 million acres they managed in thirteen southern states in 1990.[16] Public timberland REITs owned 7.7 million acres, or 4.5 percent, of timberlands in these states. The acreage

Table 5.2. Timberland area of eleven southern states by ownership and forest origin, 2010

Ownership group[a]	Planted (acres)	%	Natural (acres)	%	All forests (acres)	%
All public	2,937,131	7.4	21,334,635	16.2	24,271,766	14.2
Corporations						
Forest industry	3,517,364	8.9	3,379,648	2.6	6,895,930	4.0
Other forestry[b]	971,751	2.5	1,028,065	0.8	1,999,487	1.2
Other corporation	6,851,411	17.3	18,946,987	14.4	25,792,332	15.1
TIMOs	5,193,211	13.1	3,573,355	2.7	8,765,423	5.1
REITs	5,154,014	13.0	2,564,383	2.0	7,717,576	4.5
Individuals						
Individuals	14,523,391	36.6	78,088,582	59.5	92,586,971	54.2
Other entities[c]	492,904	1.2	2,434,260	1.9	2,926,384	1.7
Total	39,641,178	100.0	131,349,926	100.0	170,991,104	100.0

Source: Zhang, Butler, and Nagubadi (2012). They estimated these results from US Forest Service, "FIA Data Mart, FIADB Version 4.0," http://apps.fs.fed.us/fiadb-downloads/datamart.html.
[a] Including AL, AR, FL, GA, KY, NC, OK, SC, TN, TX, and VA. Data for KY, SC, and TN were for the year 2009. No data were available for LA or MS.
[b] Including forestry consultants, logging firms, and incorporated tree farmers.
[c] Including nongovernmental conservation organizations, unincorporated partnerships, associations and clubs, and tribes.

of TIMOs and timberland REITs combined reached 16.5 million acres, or 9.6 percent. Industrial timberland ownership was about 4.0 percent. Another large share of timberlands—some 26 million acres, or 15 percent—belonged to the "other" category. These were nonforestry corporations whose many businesses included utilities, mining, family farms, and real estate.

This same study also shows that TIMOs and REITs managed or owned disproportionately more (about 10.3 million acres, or 26 percent) of the planted forests in these states and that most of the forests managed or owned by TIMOs and REITs were softwoods (table 5.3). Further, the per-acre inventory of merchantable volume of growing stock on timberlands managed or owned by TIMOs and REITs was lower than that owned by individuals and other corporate owners. This indicates that forests owned or managed by TIMOs and REITs were more intensively managed and harvested than other forests. Forest industry owners had forest types similar to those of TIMOs and REITs. In contrast, individuals, other corporations, conservation organizations, and public owners had hardwoods as the dominant forest type (table 5.3).

More importantly, this study shows that timber growth exceeds removal for TIMOs and REITs as well as other owners (table 5.4). Nonetheless, TIMOs

Table 5.3. Merchantable volume of growing stock on timberland in eleven southern states by ownership and forest type

Ownership group*	Softwoods (million cubic feet)	Mixed (million cubic feet)	Hardwood (million cubic feet)	Nonstocked & other (million cubic feet)	Total (million cubic feet)	Per acre (cubic feet)
All public	13,935	5,043	25,529	544	45,052	1,856
Corporations						
Forest industry	5,251	365	3,712	3	9,330	1,353
Other forestry	1,338	140	784		2,262	1,131
Other corporations	12,993	3,196	19,936	158	36,284	1,406
TIMOs	6,717	685	3,316	14	10,732	1,224
REITs	5,769	822	2,296		8,887	1,151
Individuals						
Individuals	36,488	13,896	84,110	818	135,312	1,461
Other entities	1,121	522	3,187	135	4,966	1,697
Total	83,612	24,669	142,871	1,673	252,824	1,479

*See notes under table 5.2.

Table 5.4. Per acre net annual growth and removal (in cubic feet) by forest type and ownership group in eleven southern states

Ownership group*	Softwoods		Hardwoods		Total	
	Growth	Removal	Growth	Removal	Growth	Removal
All public	53.3**	17.8	50.9	2.8	51.8	8.4
Corporate						
Forest industry	133.4	74.4	35.6	39.4	97.7	61.7
Other forestry	123.8	75.4	30.4	4.8	84.4	45.6
Other corporations	110.8	72.0	37.7	35.1	67.1	50.0
TIMOs	129.2	81.7	34.5	63.0	100.7	76.1
REITs	130.4	89.6	38.8	74.9	103.6	85.3
Individual						
Individuals	114.1	44.0	42.0	30.4	63.3	34.4
Entities	120.4	90.7	57.1	57.6	76.5	67.7
Average	109.9	56.0	42.3	29.6	68.1	39.7

* See notes under table 5.2.
** All in cubic feet.

Table 5.5. Probability of reforestation within two
inventory cycles by ownership in the US South

Ownership	Estimated probability (%)
Forest industry	80
TIMOs and REITs	84
All other ownerships	69
Average of all ownerships	75

Source: Sun, Zhang, and Butler (2015).

and REITs as well as forest industry owners have harvested more hardwood annually than its net growth. This suggests that these managers and owners either converted some hardwood forests to softwood forests, or simply anticipated natural reproduction on these cutover hardwood sites. This phenomenon is not apparent with other landowners.

Finally, the rate of reforestation for both TIMOs and REITs was more than 80 percent, comparable to that of forest industry owners and much higher than that of all other forest landowners in the South (table 5.5). A couple of earlier studies in Mississippi also show that institutional timberland owners invested in and managed their timberlands as intensively as industrial owners.[17]

These findings suggest that institutional timberland owners do not over-harvest, do practice reforestation, and thus embrace sustainable forestry practices. At a minimum, because reforestation is a long-term investment and most institutional investors owning timberland through TIMOs often have a limited investment period of ten to fifteen years, reforestation implies that they are expecting to get an acceptable rate of return for their investment when they eventually sell their timberlands.

Forest Conservation under Institutional Timberland Ownership

Conservation organizations and environmental groups complained about the forest management practices of some forest products firms for many years before these firms divested their timberlands. The shift from industrial to institutional timberland ownership worried these conservation groups because most institutional investors want to invest in timberlands for only a number of years and have a strong financial incentive to sell their lands for the highest price. This often means that they sell portions of their holdings for HBU purposes, which can lead to land use change as well as fragmentation and parcelization of the larger forest landscape.[18]

Thus, conservation groups try to raise money and buy portions of timberlands that have high ecological value. A good example of this was the 1998 purchase of 185,000 acres of forestlands by The Nature Conservancy from International Paper in Maine. Another example was the Trust for Public Land's purchase of 2,500 acres of lands surrounding Lindbergh Lake in Montana, which was supported by the federal Land and Water Conservation Fund and revenues from the state of Montana.[19] In 1999, the Conservation Fund, which started in 1985 and aimed to provide capital to finance conservation, purchased 296,000 acres in New York, Vermont, and New Hampshire from Champion International. This purchase protected wildlife and working forests for generations and served as the concept for the fund's eventual Working Forest Fund. When International Paper Company announced the sale of 5.4 million acres of timberlands in 2005, the Conservation Fund and The Nature Conservancy bought some 300,000 acres of land with high conservation value in the southern United States and Wisconsin.[20] Thus, this shift in industrial timberland ownership gave conservation groups and government agencies a chance to buy some of these lands and put them to conservation use permanently.

Conservation and environmental groups soon discovered that raising the necessary funding in a relatively short period was a challenge. Furthermore, it was simply impossible to buy all lands on the market that they wanted to conserve. Thus, they prioritized their purchasing targets and increasingly used conservation easements as a tool to conserve forest resources.

Conservation and environmental groups even used conservation-minded TIMOs to help finance the purchase or part of the purchase and then set up conservation easements. The Forestland Group was among the first TIMOs involved in such transactions; it purchased 114,000 acres of forestlands in New York (out of 300,000 acres in New York, Vermont, and New Hampshire) and put them in an easement in return for lowering property taxes. The rest of the lands in New York went to the state of New York, and the lands in other states went to the Conservation Fund.

Most of these working forest conservation easements were sold to conservation NGOs and governments and thus allowed a handful of TIMOs to earn competitive returns on timberland investment while preventing land use change and forest fragmentation and parcelization. The easements could apply to all forestlands actively managed for their associated goods and services, including timber resources, recreational opportunities, and other ecosystem services such as carbon sequestration and biodiversity conservation. Selling conservation easements is essentially selling a "bundle of rights" for recreation, water, mineral, timber, and development associated with private timberlands. The aforementioned sale of recreational rights from Forest Capital Partners to the state of Minnesota is one example.

To date, The Lyme Timber Company may have sold more conservation easements than any other TIMO in terms of acreage under management. The Lyme Timber Company is often chosen as a partner by conservation interests because it can bring private capital to a deal up front and hold the land while its conservation partners assemble public and private money to purchase the easement on it. In doing so, it can maintain a steady cash flow, realize a return on its investment from the easement sale, and eventually sell the easement-encumbered land for a reduced price to another TIMO or timberland REIT.[21] In some cases, the conservation easement pays some 50 percent of the land purchase price.

Often accompanied by easements, some previously closed areas were opened for public recreation. Some twenty federal programs alone supported the purchase and/or establishment of easements for timberlands with high conservation value. Many states had similar programs as well. According to the Trust for Public Land, the federal funds used to purchase land for conservation or buy conservation easements amounted to $14 billion, and the total amount of land purchased and put under conservation easements was

Figure 5.3. Land conserved through outright purchase or conservation easements under federal programs in the US, 1998–2017

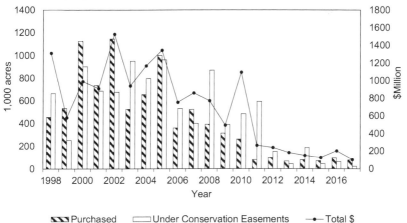

Source: The Public Land Trust, Conservation Almanac, via Jessica Welch, email message to author, September 3, 2020. The total $ includes $18.9 million spent on 3,207 acres of land that were both purchased and under conservation easements, while the acreage data include only lands that were either purchased or under conservation easements.

8.6 and 9.7 million acres, respectively, in the country in the twenty years between 1998 and 2017. Most of these lands conserved were obtained in the peak years of industrial timberland sales between 1998 and 2006 (fig. 5.3).

In short, some conservation organizations and government agencies seized the opportunities offered by large industrial timberland sales and put some of these lands into conservation via public ownership, direct ownership by conservation organizations, or institutional ownership with conservation easements. But would the new institutional owners and their TIMO managers manage the forests in an environmentally responsible way? What environmental standards would they apply without the press of forest products consumers?

Most TIMOs believe that investment objectives and environmental conservation are not contrary to each other; in fact, returns can be optimized while soil, water, air, and wildlife are protected. Certainly, sales of sensitive lands reduce management costs and increase returns. Thus, some TIMOs have even proactively reached out to environmental groups and asked for their concerns about environmental issues after they purchased lands on behalf of their investors. Having conservationists concerned about industrial timberland transformation working together with TIMOs ultimately enabled conservationists to retain some critical tracts of forestland intact and in perpetuity, forest workers to keep working in forests, and public recreationists to gain access to some previously closed forestlands. More forestlands could be protected through increased funding for existing public and private programs and partnerships between TIMOs and conservation organizations.

For institutional owners seeking to maximize returns from their timberland holdings, whatever the source, sales of "working forest" conservation easements immediately distribute cash to them mostly without unduly diminishing the ongoing cash flow from timber harvesting. Furthermore, some conservation easements buy out development benefits in one lump-sum payment and may not decrease the value of the timberland asset by as much as the payment for the easement. Recreational easements similarly purchase the recreation rights in one lump sum, avoiding the transaction costs of charging individual or club users. It is therefore not surprising that TIMOs and timberland REITs have been actively selling conservation easements, recreational easements, and hunting leases.

Additionally, some TIMOs have sold wetland banking credits, endangered species–specific conservation banking credits, and carbon credits. Empirically, the three main sources of registered forestry carbon offset projects in California have been native groups, conservation organizations, and TIMOs, even though the program may be subject to adverse selection.[22] All these payments received from environmental services enhance the returns to their investors and help conserve forest resources. Indeed, Cambridge Associates did a study on the returns of conservation-oriented TIMOs versus those with a more standard approach and concluded that the former generated higher returns between 1997 and 2014.[23] TIMOs that do conservation sales early return a lot of capital to investors early on and boost returns.

Therefore, the stewardship of institutional owners over the environmental value of forests has been at least as good as that of industrial owners and timberland REITs, if not better. These favorable outcomes have been motivated by profit motives for the sale of environmental services and pressure from institutions and their underlying participants. It just so happens that once TIMOs carry out all the conservation transactions, the internal rate of return just keeps falling the longer they own the property, which fits their institutional owners' desire to sell the property within a limited period.

Timber Supply Agreements

Timber supply agreements have existed in the forest products industry for a long time. They serve as an assurance and risk mitigation tool for both timber buyers (forest products mills) and timber sellers (log or stumpage suppliers). For timber buyers, timber supply agreements are a hedge against spot market prices. For timber sellers, they are a guaranteed market. Some integrated forest products firms use them to work through the operational impacts of timberland divestiture to TIMOs. For example, when International Paper sold 5.1 million acres of timberland in 2006, it had timber supply agreements with the timberland buyers, including RMS and FIA, for twenty years, and with TimberStar for fifty years.[24]

A typical timber supply agreement consists of a contractual obligation by a supplier to provide agreed volumes of wood to a buyer who commits to purchase them at specified contract prices. Contract prices are often based on local market prices of the previous months and quarters.[25] For most of the agreements signed by forest products firms and TIMOs, the volume is typically limited to timber harvested from the timberlands transacted between them. In practice, the agreed volume can be a specified amount of timber or a range of timber volume in a given period, or a right of first offer bestowed to the buyer or a right of first refusal bestowed to the timber seller. In the case of a right of first offer, a notice of planned harvest volume is provided to the timber buyer by the timber seller, and the buyer then decides whether to take the volume, or part of the volume, at a specified price. In the case of a right of first refusal, the buyer would get the wood only if it meets the highest offer made by other buyers.

Contract prices can be based on rolling averages, an index, or multiple indexes. Rolling (moving) averages are the average prices of the same materials supplied by others who are not under the agreement. In this case, either the timber buyer or the timber seller can keep price records and calculate the actual contract price. Indexing ties the contract price to a third-party or market index, and blended indexing ties the contract price to multiple market indexes.

Each method has its advantages and disadvantages. If either the buyer or the seller collects and reports the underlying price data or indexes, it may offer an "open book" policy, which allows the other party to inspect the data. This, along with a clear and verifiable methodology for collecting price data and reporting changes over time, gives the other party confidence that the contract price reflects market conditions and facilitates a strong timber supply relationship. A timber supply contract also defines and specifies products and timing, the process of updating and communicating prices and volumes, and the adjustment and termination of the agreement itself.

The mere existence of timber supply agreements means that they have the potential to be mutually beneficial to both timber buyers and sellers. Although some timber supply agreements have expired and not been renewed, and some timber supply agreements in retrospect have not been seen as mutually advantageous, there has not been any major dispute over timber supply agreements in the last twenty-plus years. A secure long-term timber supply is a good indicator for industrial activities that support rural communities. Also, it is possible that timber prices could be a little more volatile if all timber purchased by forest products mills came from arm's-length transactions.

Industrial Support for Forestry Research and Development

When they owned timberlands, forest products firms in the United States traditionally supported forestry research and development (R&D) activities mainly in two ways. First, some of them, such as International Paper, Weyerhaeuser, Georgia-Pacific, and Westvaco (MeadWestvaco later), had their own R&D departments focusing on improving forest productivity on

Figure 5.4. Industrial membership in the Auburn University Southern Forest Nursery Management Cooperative and North Carolina State University Cooperative Tree Improvement Program, 1985–2020

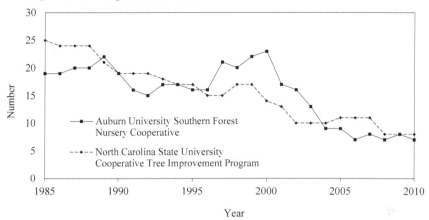

Sources: Auburn University Southern Forest Nursery Management Cooperative annual reports (various years) and "TIP Annual Reports," NSCU Tree Improvement Program, accessed November 30, 2020, https://www.treeimprovement.org/annual-reports.

their own lands. More importantly, most of them were members of, and provided financial support to, various forestry research cooperatives led by forestry schools across the country. Through these cooperatives, forest products firms collectively identified critical research needs, selected research topics, and secured leading-edge research results quickly.

TIMO (and to a lesser extent, timberland REIT) expenditures on forestry R&D have been far lower than those of their prior industrial counterparts. Basically, no TIMO funds independent R&D on its own hook, or via the companies it manages. Very few TIMOs participate in university forestry research cooperatives. For example, no TIMO has been a member of the Auburn University Southern Forest Nursery Management Cooperative, and only one TIMO has participated in the North Carolina State University Cooperative Tree Improvement Program since 2000. As a result, the industrial membership of both cooperatives declined after the late 1990s and early 2000s when industrial mergers and acquisitions as well as sales of industrial timberlands to institutions took place (fig. 5.4). Similar declining industrial participation has occurred in most other forestry research cooperatives around the country. Consequently, a few research cooperatives have dissolved, and those that remain active have had to raise membership fees, charge service fees, and cut operating expenses.

This lack of investment in R&D by TIMOs could be explained by the relatively short duration of most institutional timberland funds, the time needed to produce research results (up to a couple of years), the fact that most research is for the public good (most results would be published anyhow), and the potential for TIMOs to buy the results of research embodied in plant materials from ArborGen, for example. Stated in another way, TIMOs have said that they cannot justify and sell R&D needs to their institutional investors, even though many research cooperatives have been courting them to become members for some time.

Thus, although part of the slack in R&D that used to be funded by the forest products industry has been picked up by the remaining university-led cooperatives and private firms such as ArborGen, the rise in institutional timberland ownership has had a negative impact on forestry R&D overall in the country.

Summary of TIMOs

Currently, there are about thirty active TIMOs in the United States, although the top ten represent over 95 percent of the TIMO business. All these TIMOs advise their investors and try to add value by working on the demand side of timberland sales as well as the supply side in terms of acquisition, portfolio management, forest management, income generation, and ultimately timberland disposal. They all tout the same beneficial attributes of investing in timberlands. Their strategies for identifying profitable investments explore various market inefficiencies, such as

- potential acquisitions/deals obtained through a well-developed network of contacts;
- arbitrage opportunities between sellers and buyers who value timberland differently because of differences in tax and valuation metrics, accounting rules, and needs;
- arbitrage opportunities for forests of different ages;
- opportunities in various species mixes and in places where governments support businesses in less-developed communities (through low-interest loans);
- regional (and international) opportunities;
- wholesale discounts;
- imperfect information;
- value added by applying technical forest management expertise; and

- buying low and selling high by paying close attention to, and having superior knowledge of, timber markets, conservation easements, wetland mitigation and conservation banks, and carbon markets.

Most TIMOs diversify within the timberland asset class by investing across a range of merchantability classes and secure value by investing only in lands with productive soils, road access, and a history of good management. Most acquisitions are made for cash, although leverage has been used.

As TIMOs compete and strive to add value, they help improve the efficiency of timber and timberland markets. But as far as operations and business models are concerned, most TIMOs are very similar. TIMOs that are subsidiaries of insurance companies and banks can have more corporate support but are limited by the strategies and needs of their parent companies. The disappearance of Travelers as a TIMO had more to do with the needs of Travelers than with the TIMO business itself.

Privately held TIMOs, on the other hand, are independent firms engaged in the investment advisory business and registered with the Securities and Exchange Commission (SEC) and/or state securities authorities. Privately held TIMOs often start from the ground without much corporate support, but they may have more flexibility and potentially lower management fees.

TIMOs compete among themselves as well as with other entities, such as publicly traded timberland REITs, which are the subject of the next chapter.

CHAPTER 6
Timberland REITS

In this chapter I discuss timberland MLPs and their conversion to timberland REITs in the United States, their respective advantages and disadvantages versus other corporate structures, their attractiveness to institutional investors, and their operations and management. I also reveal the financial attributes of timberland REITs and compare the financial performance of TIMOs and timberland REITs (including their predecessors—timberland MLPs) with other financial assets in the thirty-four years between 1987 and 2020, with an emphasis on the last decade. Finally, I attempt to answer the question of which investment vehicle was, is, and will be better for institutional investors.

The key points of this chapter are three. First, tax policy has an enormous impact on industrial structure, for better or worse. Better in that public investors finally got relatively undiluted exposure to timberland; worse in that the Brazil-style benefits of integrated firms were eliminated in the United States. Second, returns from public and private equity timberland may well differ. But the differences might be due more to construct and leverage than to underlying returns. By "construct" I mean that private timberland equity represented in the NCREIF Timberland Index likely understates return volatility due to appraisals. By "leverage" I mean that all publicly traded timberland REITs use various degrees of debt, which private equity rarely uses. All else being equal, leverage often leads to greater returns as well as higher volatility. Finally, although private and public real estate equity tends to keep its value in line with fundamentals—if one becomes expensive, development occurs to bring value back in line with the other—this does not happen often with timberlands. This situation may have allowed some savvy investors to purposefully stick with one investment vehicle for a while and secure adequate risk-adjusted returns from timberland investment in the past few decades.

More Timberland MLPs Emerging

As noted in chapter 3, US forest products firms that faced pressure to monetize their timberlands and were subject to takeover bids in the early 1980s chose to sell some of their timberlands outright, set up master limited partnerships (MLPs), or treat their timberlands as an independent profit center. The first MLP, the Timber Realization Company, created by Masonite Corporation in July 1982, was a liquidating limited partnership whose purpose was to sell the property and business it owned and return each partner a share of the remaining money. It ceased operations in 1987 after accomplishing its mission.

Following its IPO in December 1984, IP Timberlands became the first publicly traded MLP on the New York Stock Exchange. Both Rayonier and Pope Resources went public in October 1985. These three MLPs were still tied to their respective parent companies. F. Christian Zinkhan concluded that shareholders of these companies benefited significantly around the dates of these spin-off announcements.[1]

The appeal of MLPs was threefold. First, investors would get public market liquidity and exposure to pure timberland returns without the returns and capital reinvestment requirements of forest products manufacturing. Second, and very importantly, MLPs were beneficial as "flow-through vehicles" with single-level taxation, even on ordinary income. Additionally, timber sale income was capital gains income, which was then taxed at a much lower rate than the ordinary income generated by manufacturing activities.

Under the MLP partnership agreements, the unit holders of IP Timberlands LP were entitled to receive 95 percent of the respective partnership's net cash flow for fifteen years and 5 percent afterward. In addition, International Paper sold only 20 percent of the units in the MLP and retained the rest. Similarly, Rayonier held 75 percent of the units of Rayonier Timberlands LP, whose life was fifteen years. As noted earlier, Rayonier Timberlands LP was reintegrated with Rayonier in February 1998, and IP Timberlands LP was absorbed by International Paper Company in April 1998.

In June 1989, Plum Creek Timber Company went public as a timberland MLP. Plum Creek was a fully owned subsidiary of Burlington Northern, which was a large asset-holding company with various railroad, timberland, petroleum, mineral, and manufacturing assets. The timberlands were originally part of the checkerboard pattern of the nineteenth-century railroad grants, which enabled funding for construction of the transcontinental railroad. Burlington Northern's management realized that the price of its company stock did not account for the combined value of its diverse holdings. So, in 1988, Burlington Northern spun off its resource businesses, including timberlands, mineral rights, and petroleum resources, into a holding company called Burlington Resources. It then sold off subsidiaries, including Plum

Creek Timber Company, which had timberlands and a few forest products manufacturing plants. At the suggestion of its investment bank, Plum Creek chose to spin off as an MLP for tax advantages.

Burlington Resources thus became the general and controlling partner, and the initial public offering raised $256 million through the sale of depository units (which are like shares of corporate stocks), representing limited partner interests, at an initial cost of $20.50 each. With that capital and $325 million in loans, Plum Creek bought 1.4 million acres of timberland, half of which were in Montana and the rest in Washington, Oregon, and Idaho, as well as manufacturing facilities from its parent company, at fair market value.[2] Should the manufacturing facilities have had no value at all, these timberlands were valued at $415 per acre at the time. In retrospect, Plum Creek got a good deal from its parent company.

The last two timberland MLPs to go public were Crown Pacific Partners LP in December 1994 and U.S. Timberlands Company LP in November 1997. Unlike other timberland MLPs that were spin-offs of major integrated forest products companies or other asset-holding companies, Crown Pacific Partners originated from a privately held timberland and forest products company called Crown Pacific Ltd., which was founded in Portland, Oregon, in 1988. In addition to money raised from its initial public offering, Crown Pacific Partners borrowed more than $500 million in the 1990s to buy timberlands. At its peak, it owned 800,000 acres of timberland, half in Oregon and the rest in Washington, Idaho, and Montana. It also had several sawmills, a wood-chip plant, and other wholesale facilities. Crown Pacific filed for Chapter 11 bankruptcy on June 30, 2003.[3] Its remaining assets, including 520,000 acres of timberland, were taken over by creditors in December 2004 and managed by Cascade Timberlands LLC.

U.S. Timberlands Company LP went public shortly after its initial purchase of 600,000 acres of timberlands from Weyerhaeuser Company in Klamath Falls, Oregon, in 1996. This purchase at the peak of timber prices, as well as poor management, caused the company to suspend dividend distributions in 2001. In 2003, it was delisted and taken private after being a public timberland MLP for only six years.

Plum Creek, the third major MLP of the 1980s, would become a timberland REIT in 1999. A few other timberland MLPs and Subchapter-C corporations followed and successfully converted themselves into timberland REITs in the 2000s. These included Rayonier Inc., which announced the change in August 2003 and officially converted in January 2004, and Weyerhaeuser

Company in 2010. As noted in chapter 5, Pope Resources remained an MLP until it was purchased by Rayonier in January 2020.

Timberland REITs Became Possible in 1997

REITs have existed since 1960 as an income-producing real estate investment entity in which the attributes of real estate and stock-based investment are combined. A REIT is a company that owns or finances income-producing real estate. Known as the "mutual funds of real estate," REITs allow anyone to invest in portfolios of large-scale properties the same way they invest in other industries—through the purchase of REIT stocks. In the same way that shareholders benefit by owning corporate stocks, the stockholders of a REIT earn a share of the income produced through real estate investment without having to go out and buy property. The REIT structure allows illiquid real estate assets to be traded as stocks.

To qualify as a REIT, a company must

- invest at least 75 percent of its total assets in real estate (asset test);
- derive at least 95 percent of its gross income and 75 percent of its net income from rents from real property, interest on mortgages financing real property, or sales of real estate (income test);
- pay at least 90 percent of its taxable income in the form of shareholder dividends each year;
- be taxable as a corporation;
- be managed by a board of directors or trustees;
- have a minimum of one hundred shareholders; and
- have no more than 50 percent of its shares held by five or fewer individuals.

Thus, REITs typically pay out most of their taxable income as dividends to shareholders. In turn, shareholders pay income taxes on those dividends. Therefore, REITs are a tax-advantaged corporate structure.

In contrast to REITs, MLPs were limiting both in what they could acquire in the private company sphere and in who could own their actual units. Even though MLPs started in the early 1980s, they were not officially sanctioned until the Tax Reform Act of 1986. Most MLPs are in the energy and natural resource sectors. A REIT is considered a financial-sector investment and is involved with only real estate. As such, REITs can have better access to capital in debt markets and thus can operate with more leverage than MLPs. The distribution requirements also differ for REITs and MLPs. REITs must pay out 90 percent of earnings in the form of dividends to their shareholders. MLPs are not required to distribute a certain percentage. Instead, MLPs declare their share of dividends on their own when they are initially set up. Distributions are also treated differently on the receiving end. While REIT distributions come with a tax liability for their investors like any other dividends, MLP distributions are often tax-free because the MLP pays corporate income taxes. However, when held within a retirement account, this income, if it is more than $1,000 for a unit holder, is considered unrelated business taxable income (UBTI). For this reason, MLPs are not ideal for individual retirement accounts (IRAs) or institutional investment.

Timberland REITs became possible after the Real Estate Investment Trust Simplification Act (REITSA) of 1997, as part of the Taxpayer Relief Act of 1997, which was enacted into law on August 5 of that year. REITSA of 1997 struck from the Internal Revenue Corporate Tax Code paragraphs 4 and 8 of Section 856(c) the provisions that described qualifying REIT income. Previously, a REIT would be disqualified if it received more than 30 percent of its annual gross income from dispositions of real property held for less than four years. These provisions made a timberland REIT impractical, since in effect they would have made it difficult or impossible for a timberland REIT to harvest timber and generate less than 30 percent of its income from properties it had held for less than four years. The 30 percent test also prevented a timberland REIT from capitalizing on opportunities to make a large profit, such as when and if it received an offer to buy a package of properties it had held for less than four years, and thus created a significant barrier for it to buy a package of properties with a view of selling a significant amount within four years.

Timber Growth Corporation (TGC) was the first company that tried to become a timberland REIT in 1997–1998. TGC had 824,000 acres of initial timberlands offered by Potlach Corporation and Anderson-Tully Company. The plan was to sell 42.9 percent of the interest through public offering in these timberlands, and Potlach would control the remaining 57.1 percent of the interest. The proceeds of the public offering would be used to pay for the timberlands contributed by Anderson-Tully Company and new acquisitions.[4] The initial public offering did not happen because the valuation (six to eight times the expected annual dividends) coming from Wall Street did not meet the expectations of either Potlach or Anderson-Tully, which were fourteen to fifteen times the expected annual dividends. This low valuation resulted partly from the dot-com bubble and partly from concerns about the initial control of TGC by Potlatch Corporation. Interestingly, this effort was led in part by three people in the TIMO business, all from Hancock—Scott Jones and Matt Donegan, who later formed Forest Capital Partners, and Courtland Washburn, who went back to Hancock.

Plum Creek was thus the first company that successfully converted to a timberland REIT. As an MLP, Plum Creek had grown from 1.4 million acres to 3 million acres in six states in its first ten years, but its units were owned primarily by retail investors. Without institutional investors, it was difficult to raise capital for additional growth. Laura Sloate of Sloate Weisman Murray & Co., who owned a position in Plum Creek for her clients, suggested to Plum Creek president and CEO Rick Holley in the summer of 1997 that the company be converted to a REIT. Intrigued, Holley and other executives explored the idea. They made an inquiry to, and got approval from, the IRS and became a REIT on July 1, 1999.[5]

The benefits of being a REIT for raising capital came almost immediately for Plum Creek. In July 2000, Plum Creek agreed to merge with The Timber Company, the timberland operating company controlled by Georgia-Pacific Corporation. The $4 billion transaction involved 4.7 million acres of timberland owned by Georgia-Pacific, which was more than Plum Creek's own previous timberland holdings, making Plum Creek the second largest private timberland owner in the United States at that time, with 7.8 million acres in nineteen states.

When Georgia-Pacific created The Timber Company as a letter stock for its timberland operations and thus separated timberlands from its forest products manufacturing operations in 1997, it also considered a spin-off. But, as A. D. Pete Correll, its chair and CEO, said, "The problem of a spinoff is that it's not very tax-efficient." Georgia-Pacific would have had to pay a hefty capital gains tax because the book value of the timberlands was low as a result of inflation, depletion of original timber, and regrowth over time. Furthermore, Correll said, "It totally takes control away from us."[6]

Nonetheless, the separation of timberlands from its paper manufacturing business was a departure from the traditional timberland ownership and vertical operating philosophy of forest products manufacturers. By selling The Timber Company three years later, Georgia-Pacific demonstrated that even one of the largest forest products firms did not have to own timberlands to stay competitive. Georgia-Pacific secured a ten-year timber supply agreement with Plum Creek, although it acquired about 80 percent of its raw materials on the open market at the time of the sale. This timber supply agreement was not renewed when it expired in 2011.

The transaction with Plum Creek was tax-free because the acquisition corporation, Plum Creek, was smaller than The Timber Company, and a little-known IRS rule called the Reverse Morris Trust was used. The Reverse Morris Trust is a tax-avoidance strategy used by corporations wanting to dispose of unwanted assets while avoiding taxes on any gains from those assets. The Reverse Morris Trust starts with a parent company (Georgia-Pacific in this case) looking to sell assets to a smaller external company. The parent company then creates a subsidiary (The Timber Company in this case). That subsidiary and the smaller external company (Plum Creek) merge and create another company (also named Plum Creek) unrelated to the parent company (Georgia-Pacific). The unrelated company then issues shares to the shareholders of the original parent company. If those shareholders control over 50 percent of the voting right and market value in the unrelated company, the Reverse Morris Trust is complete. The parent company (Georgia-Pacific) effectively transfers the assets to the shareholders of the smaller external company (Plum Creek) tax-free.

Other Timberland REITs

When Plum Creek successfully became a REIT, some other timberland MLPs and forest products companies also considered converting to REITs. The problem was passing the asset and income tests to qualify for REIT status. Firms with substantial manufacturing assets would not be able to pass these tests.

A piece of legislation called the REIT Modernization Act (RMA) of 2001 came to the rescue. The RMA of 2001 enables a "taxable REIT subsidiary" (TRS) to exist within a REIT that can provide services. A TRS is not subject to the REIT asset, income, and distribution requirements, nor are its assets, liabilities, or incomes treated as the assets, liabilities, or incomes of a REIT for purposes of each of the above REIT qualification tests. A TRS is taxed as a C corporation.[7]

Thus, in theory, most forest products firms with timberlands and manufacturing assets registered as Subchapter-C corporations could become REITs as Subchapter-S corporations, using their timberlands and setting up their manufacturing assets under a TRS. In this way, the double taxation on income derived from their timberlands would be eliminated.

Four other forest products firms—Rayonier, Potlatch Corporation, Longview Fibre Company, and Weyerhaeuser Company—had taken this route and converted to REITs before another piece of legislation, the Protecting

Americans from Tax Hikes Act of 2015, banned any conversion from C corporations to REITs.

Rayonier became a timberland REIT in a fashion similar to Plum Creek. Rayonier was founded as Rayonier Pulp and Paper Company in Shelton, Washington, in 1926. It was acquired by International Telephone and Telegraph Corporation (ITT) and became ITT Rayonier in 1968. In 1985, ITT Rayonier created Rayonier Timberlands, an MLP, for its 1.2 million acres of timberlands. In 1994, Rayonier once again became an independent public company in a spin-off from ITT Rayonier. In February 1998, Rayonier bought back all outstanding units of Rayonier Timberlands and reintegrated timberlands with its manufacturing business, becoming an integrated Subchapter-C corporation. In January 2004, it converted itself to a REIT. In June 2014, Rayonier spun off its performance fiber business into an independent, publicly traded company named Rayonier Advanced Materials Inc. and thus made itself a pure timberland company.

Potlatch Corporation became a REIT on January 1, 2006. Potlatch had 1.5 million acres of timberland. Because REIT tax rules required that the company derive most of its income from investment in real estate, Potlatch transferred almost all its manufacturing facility assets to a wholly owned TRS prior to the conversion.

Also in January 2006, the Longview Fibre Company of Longview, Washington, became the fourth timberland REIT in the United States. In February 2008, Brookfield Asset Management of Toronto (a TIMO) bought Longview Fibre for $2.15 billion, including assumed debt, and took it private. In June 2013, Brookfield sold Longview Fibre's paper and packaging business for $1.03 billion to KapStone Paper & Packaging Corp., an Illinois company. In the following month, it sold the 645,000 acres of timberlands that it had acquired from Longview in Washington and Oregon to Weyerhaeuser Company for $2.65 billion, which, as noted in chapter 1, was priced at $4,108 per acre. This deal generated an annual return of nearly 11 percent in capital appreciation alone, more than tripling the annual average return in the NCREIF Timberland Index in the same period.

Also in 2006, Wells Real Estate Funds—a private venture capital fund—sponsored a private REIT called Wells Timberland REIT Inc. and offered its shares mostly to retail timberland investors through brokers and financial planners, with a minimum investment of $5,000. The proceeds were used to invest in timberland properties. In December 2013, Wells Timberland became a public REIT under the name CatchMark Timber Trust.

As the smallest publicly traded timberland REIT, with 404,000 acres of wholly owned timberlands in the United States, CatchMark significantly boosted its asset base in 2018 by acquiring an interest in 1.1 million acres of prime East Texas timberlands through a joint venture, TexMark Timber Treasury LP—known as Triple T Timberlands. This joint venture was with a consortium of institutional investors, including BTG Pactual Timberland Investment Group, Highland Capital Management, Medley Management Inc., and British Columbia Investment Management Corporation. CatchMark provided an investment of $200 million in the $1.39 billion transaction initiated by Campbell Global representing the general and limited partnership interests of the entities that had previously owned the timberland assets.[8]

* * *

The last C corporation to become a public timberland REIT was Weyerhaeuser Company in 2010. Despite the burden of proof for qualifying as a REIT, the benefit of eliminating double taxation was too attractive for Weyerhaeuser, which had its roots in timberland investment and had timberlands as its most valuable assets. As noted earlier, a report by Lehman Brothers in 2006 shows that the difference in timber income between a double-taxed C corporation and a tax-advantaged ownership structure can be as high as 39 percent.[9] Thus, selling timberlands outright and converting to REITs were encouraged and fashionable for forest products companies in the mid-2000s.[10]

Because of TRS, the installment sales technique, and the Reverse Morris Trust, most vertically integrated forest products companies had either sold their timberlands outright or converted themselves to REITs by 2006. Weyerhaeuser was under pressure to make such a conversion as well. It postponed its conversion until the Timber Revitalization and Economic Enhancement Act (TREE Act) of 2008 became law.

Between 2005 and 2008, the forest industry, led by Weyerhaeuser, lobbied Congress to allow forest products firms to pay the same level of tax as other timberland owners. Weyerhaeuser hoped the TREE Act of 2008 would subject all timber income to the same federal income tax rates irrespective of ownership.[11] Had this lobbying effort succeeded, it would have leveled the playing field for all forest landowners as far as federal income tax was concerned.

However, the TREE Act of 2008 offered only one-year temporary tax relief by allowing forest products companies to pay the same level of taxes as other landowners—15 percent for gains from timber they had held for more than fifteen years. With no other viable option to level the playing field in terms of tax liability, Weyerhaeuser, the last large, traditional, vertically integrated forest products firm, announced on December 15, 2009, that it would convert itself to a timberland REIT in January 2010. In doing so, it had to sell some forest products manufacturing assets and put its remaining non–real estate business in a TRS.

In 2016, Weyerhaeuser purchased Plum Creek, becoming the owner of nearly thirteen million acres of timberland in the United States. In February 2018, Potlatch Corporation acquired Deltic Timber Corporation and changed the name of the combined company to PotlatchDeltic Corporation.[12]

* * *

Our list of timberland REITs should also include a Canadian firm—Acadian Timber Corporation. As an income trust, Acadian Timber currently holds three hundred thousand acres of timberland in the United States, in addition to its substantive timberland holdings and Crown lands under management in Canada. More importantly, under Canadian law, it operates just like a timberland REIT in the United States: it is a corporate vehicle without corporate tax, passing through income to its shareholders, who are mostly income seekers. Acadian Timber is also listed in the OTC (over-the-counter) markets in the United States.

Acadian Timber started with a public offering of the Acadian Timber Income Fund (a mutual fund) by Brookfield Asset Management Inc. that consisted of approximately 765,000 acres of freehold timberlands and related assets in the province of New Brunswick from Fraser Papers Inc., and 311,000 acres of freehold timberlands in Maine in January 2005. On January 1, 2010, the Acadian Timber Income Fund was converted to Acadian Timber Corporation as an income trust under the Canada Business Corporations Act. The management and trustees of the fund became the management and directors of the corporation.

This conversion was in advance of Canadian federal income tax legislation relating to specified investment flow through trusts, which was announced on October 31, 2006, and implemented in January 2011. The new Canadian legislation contained "normal growth" and "undue expansion" restrictions that could have limited the fund's ability to consider strategic acquisitions. Thus, this conversion was intended for greater access to capital and more expansion opportunities for Acadian Timber. In this case, tax policy had a significant impact on industrial structure in Canada, as in the United States.

* * *

As table 1.1 shows, at the end of 2020, there were four publicly traded REITs in the United States: Weyerhaeuser Company, Rayonier Inc., PotlatchDeltic Corporation, and CatchMark Timberland Trust Inc.; and one in Canada: Acadian Timber Corporation. Collectively, they owned own some 16.6 million acres of timberland in the United States.

Investing in Timberland REITs versus TIMOs

The coexistence of private equity through TIMOs and publicly traded timberland REITs provides alternatives for institutional investors to invest in timberlands. At the very minimum, investors who invest in one can look at their "opportunity costs" ex post if they invest in the other. Jack Lutz of Forest Research Group in Alton, Maine, refers to TIMOs and REITs as pizza and beer or tea and coffee. Investors can consume one without the other or both, depending on their preference. This analogy implies that they are complementary.

Yet timberland investments through TIMOs have so many quirks that the asset class makes sense only for institutional and private investors who have a long-term perspective and want more direct control and some economy of scale. While TIMOs cater largely to institutions, they often accept individuals investing a minimum of between $1 million and $5 million in commingled funds. Some TIMOs allow individuals to pool money into a single account to meet their minimum requirements. But that is still a lot of money to devote to an illiquid investment that cannot be unloaded easily and must be held for eight to ten years just to begin to pay off. Here I do not count the lucky few that were able to flip a large amount of timberlands and generate hefty returns in just a few months or a couple of years, as these cases are very rare.

For both small and large investors who want to buy timberlands as a highly liquid asset, the best option might be to buy the shares of a publicly traded timberland REIT. Timberland REITs generally offer decent dividend yields and are diversified across multiple regions and species. The drawbacks are that timberland REITs are susceptible to timber-industry dynamics and macroeconomic conditions: they all tanked amid the great financial crisis. Also, timberland REITs might be pressured to sell trees in a weak market to maintain the healthy payouts that investors expect, cutting into future returns. On the other hand, a TIMO might not sell timber in a weak market if it thinks prices will recover. If prices do not recover for a long period, the

TIMO will have a low return. In some unusual circumstances, timber that could have been harvested around 2006, when sawtimber prices were $45 per ton, was sold at $25 per ton in the last few years (fig. 2.1).

The other difference is in valuation metrics. The stock of a timberland REIT is often valued based on its earnings multiples, such as its price-to-earnings ratio, although its future earnings are estimated for three to five years and assumed with constant growth rates afterward. Owning timberland REITs is more for quarterly dividends.[13] On the other hand, investing through TIMOs is more for total returns.

Investors are interested mostly in the returns and risk of their investments. In the next section, I compare the financial performance of TIMOs and timberland REITs in the last few decades.

Assessing the Financial Performance of TIMOs and Timberland REITs

I have constructed two dynamic portfolios of public timberland REITs (including their predecessors, MLPs) in the United States. I have then evaluated the financial performance of these portfolios compared with the NCREIF Timberland Index and the broad financial markets represented by the S&P 500 between 1987 and 2020.

My first dynamic portfolio of US publicly traded timberland MLPs and timberland REITs includes

- IP Timberlands LP (January 1987–March 1998)
- Rayonier Timberlands LP (January 1987–January 1998)
- Plum Creek (July 1989–March 2016)
- Pope Resources LP (January 1987–December 2019)
- The Timber Company (January 1998–October 2001)
- Rayonier Inc. (January 2004–December 2020)
- Potlatch (January 2006–January 2018)
- PotlatchDeltic (February 2018–December 2020)
- Weyerhaeuser (January 2010–December 2020)
- CatchMark Timber Trust (January 2014–December 2020)

My second dynamic portfolio includes all firms in our first dynamic portfolio plus two other timberland MLPs and one timberland REIT, which existed for a short time during the study period:

- Crown Pacific Partners LP (January 1995–February 2003)
- U.S. Timberlands LP (December 1997–November 2002)
- Longview Fibre (January 2006–March 2007)

As noted earlier, both Crown Pacific Partners and U.S. Timberlands went bankrupt, and Longview Fibre was bought by Brookfield Asset Management in 2007. This second portfolio eliminates the possible survival bias that may exist in the first portfolio. Yet the results of these two portfolios are very close because adding these three firms affected the returns in only a limited number of years, and the capitalization of these three firms was small. Thus, unless I explicitly mention both portfolios, I use the second portfolio to represent timberland REITs in the following discussion.

It is important to note that the returns represented by the NCREIF Timberland Index are the gross of investment management fees and other "fund-level" expenses. To rectify this shortcoming, NCREIF introduced in 2012 its Timberland Fund and Separate Account Index (TFI), which takes these factors, along with leverage, into account. Based on an analysis of the differences between the NCREIF Timberland Index and TFI returns, Chung-Hong Fu of Timberland Investment Resources found that fees consumed about 1 percent of the return for a typical investor.[14] On the other hand, the returns from timberland REIT portfolios are the net of all management fees.

Table 6.1 presents the average annual rates of return for the two portfolios, the NCREIF Timberland Index, and the S&P 500, as well as their standard deviations. It shows that the two timberland REIT portfolios outperformed the NCREIF Timberland Index with much higher (80 percent higher) volatility. Timberland REIT portfolio no. 2 performed only slightly worse than timberland REIT portfolio no. 1. Both portfolios had an average annual rate of return of more than 15 percent, substantially higher than that of the NCREIF Timberland Index and S&P 500 but with higher volatility.

Figure 6.1 shows the cumulative value of timberland REITs (again represented by portfolio no. 2), along with that of the NCREIF Timberland Index and the S&P 500 in the study period using quarterly return data. It shows that a dollar invested in the S&P 500 at the end of 1986 was, with dividends included, worth $32.70 at the end of 2020, at an average compound annual rate of return of about 10.8 percent. The same dollar invested in the NCREIF Timberland Index then was worth nearly $32.50 at the end of 2020, at an average compounded annual rate of return of 10.78 percent. On the other hand, timberland REITs had an average compounded annual rate of return of 13.44 percent.[15] Although the difference in the average compounded rates of return between timberland REITs and the NCREIF Timberland Index was merely 2.6 percent per year, the power of compounding over the thirty-four

Table 6.1. Average annual rates of return and standard deviation for two timberland REIT portfolios, the NCREIF Timberland Index, and the S&P 500, 1987–2020

	Timberland REIT portfolio no. 1 (%)	Timberland REIT portfolio no. 2 (%)	NCREIF Timberland Index (%)	S&P 500 (%)
Average annual rate of return	15.60	15.31	11.26	12.22
Standard deviation	19.23	19.08	10.66	16.84

Note: See text for the composition of timberland REIT portfolios.

Figure 6.1. Cumulative total returns for timberland REIT portfolio no. 2, NCREIF Timberland Index, and S&P 500, 1987–2020

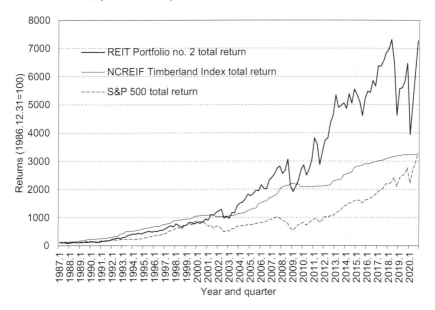

years made the value of the former investment more than twice that of the latter, with nearly twice the volatility.

Figure 6.1 also shows that the NCREIF Timberland Index outperformed the dynamic timberland REIT portfolio before 2001 and lagged in most years afterward. This is to say that in the first fifteen years of our study period, private timberland investments outperformed public timberland investments, and the situation reversed afterward.

These results may be related to the following facts: (1) timberland REITs did not exist until August 1999, and timberland MLPs were not attractive to many institutional investors and underperformed initially; (2) two of the

largest timberland MLPs—IP Timberlands and Rayonier Timberlands—
were largely controlled by their parent companies and had limited duration; (3) The Timber Company was controlled by Georgia-Pacific; and (4) all
remaining timberland MLPs were too small to take advantage of the inefficiency in timber and timberland markets or to attract institutional investors.
Conversely, private timberland investments through TIMOs were quick to
exploit and capitalize on the market inefficiency in the first fifteen years of
institutional timberland investment. Perhaps as the market for timberland
investment developed, timberland REITs became a better option but with a
higher level of risk.

The cumulative return of timberland REITs caught up with that of the
NCREIF Timberland Index in 2001, and these two series crossed each other
again only in 2009. We see three phases of rapid growth in this timberland
REIT portfolio. The first phase, between 2000 and 2002, was probably
related to two things: Plum Creek becoming the first timberland REIT, and
the merger between Plum Creek and The Timber Company.

The second phase of rapid growth in the timberland REIT portfolio was
between 2003 and 2007. In this period, three firms—Rayonier, Potlatch, and
Longview Fibre—successfully converted themselves into REITs. The great
financial crisis of 2008–2009 led to a sharp decline in the value of the timberland REIT portfolio.

The third phase was between 2009 and 2018. This was perhaps related
to the conversion of Weyerhaeuser into a timberland REIT and its purchase
of Plum Creek in early 2016. In addition, the record low interest rates in this
period allowed timberland REITs to retire their high-interest debt and refi-
nance their operations.

Again, these absolute returns shown in figure 6.1 should be considered
along with the accompanying risk. As table 6.1 shows, the risk, measured by
volatility, of timberland REITs was 10 percent higher than that of the S&P
500 and about 80 percent higher than that of the NCREIF Timberland Index.

* * *

Now let us study the return and risk profiles of these timberland REITs using
two popular models in the finance literature—the Capital Asset Pricing
Model (CAPM), which is described in chapter 3, and the Fama-French three-
factor model.

The Fama-French three-factor model was developed based on empirical evidence that small-sized stocks outperform large-sized stocks, and value

(high book/market ratio) stocks outperform growth (low book/market ratio) stocks on average. Thus, these extra two factors are added to CAPM to adjust for risk for investment i:[16]

$$E[R_i] - R_f = b_{RMRF, i} E[R_{RMRF}] + b_{SMB, i} E[R_{SMB}] + b_{HML, i} E[R_{HML}] \qquad (5.1)$$

where:

R_{RMRF} = $R_M - R_f$ is the same market factor as in CAPM, representing the market risk premium;

R_{SMB} = $R_{small} - R_{big}$ is the size factor, representing the return difference between a portfolio of small stocks and a portfolio of large stocks (SMB stands for "small minus big");

R_{HML} = $R_{highBM} - R_{lowBM}$ is the book-to-market factor, representing the return difference between a portfolio of high book-to-market stocks and a portfolio of low book-to-market stocks (HML stands for "high minus low"); and

βs are called factor loadings, representing each asset's sensitivity to these factors.

When estimating the Fama-French three-factor model, I have used ex post realized returns, as in the case of CAPM, and an intercept (α_i) is added to capture the abnormal performance:

$$R_i - R_f = a_i + b_{RMRF, i} R_{RMRF} + b_{SMB, i} R_{SMB} + b_{HML, i} R_{HML} + \varepsilon_i \qquad (5.2)$$

Here, R_f is represented by the one-month US Treasury bond yield, and R_M is represented by the value-weighted returns of all stocks listed on the New York Stock Exchange (NYSE), AMEX, and NASDAQ. All data—including R_{RMRF}, R_{SMB}, and R_{HML}—are quarterly and collected from the Center for Research in Security Prices at Wharton Research Data Services.[17]

Table 6.2 presents the estimated results of CAPM on the quarterly returns of two dynamic timberland REIT portfolios and of the NCREIF Timberland Index. A significant positive α is found in the two portfolios and the NCREIF Timberland Index, suggesting that public- and private-equity timberland investments both have a risk-adjusted excess return. The magnitude of this excess return for timberland REIT portfolio no. 2 is around 7.82 percent [(1 + 0.019%)^4 – 1 = 7.82%], and that for the NCREIF Timberland Index is 10.81 percent per year.

While the finding of excess returns for the NCREIF Timberland Index is consistent with the literature, the existence of excess return for the timberland REITs is revealed here for the first time, as far as I know.[18] Market βs for the two timberland REIT portfolios are significantly different from 0 and smaller than 1, suggesting that timberland investment through REITs

Table 6.2. Estimated results of the Capital Asset Pricing Model on timberland REIT portfolios and the NCREIF Timberland Index using quarterly data, 1987–2020

	Timberland REIT portfolio no. 1	Timberland REIT portfolio no. 2	NCREIF Timber-land Index
Alpha (α)	0.020** (0.008)	0.019** (0.003)	0.026** (0.000)
Beta (β)	0.852** (0.000)	0.857** (0.000)	-0.009 (0.808)
R^2	0.442	0.456	0.000
R^2 adjusted	0.438	0.452	-0.007
Durbin-Watson statistic	1.939	1.973	1.662
F statistic	106.20	112.20	0.059

** Significant at the 1% level.
Note: Numbers in parentheses are *p* values. See text for the composition of timberland REIT portfolios.

has lower systematic risk than the stock market as a whole. This finding is similar to the previously estimated β of 0.52 for a timberland MLP portfolio of IP Timberlands, Rayonier Timberlands, Plum Creek, and Pope Resources between 1987 and 1997 (table 6.3). The β for the NCREIF Timberland Index is insignificant. This finding is consistent with the literature listed in table 6.3 as well.

On the other hand, the Fama-French three-factor model fits the returns of the two timberland REIT portfolios much better, as implied by the higher R^2 values (table 6.4). While the estimated β for market factor β_{RMRF} changes little from CAPM, the highly significant coefficients in β_{SMB} and β_{HMI} imply that these two extra factors add additional explanatory power in pricing

Table 6.3. Comparison of Capital Asset Pricing Model results on timberland REIT portfolios and the NCREIF Timberland Index using quarterly data

	Timberland MLP portfolio	Dynamic timber-land portfolio	NCREIF Timberland Index	
Alpha (α)	0.020	0.0057	0.042[a]	0.0234[a]
Beta (β)	0.52[a]	0.95[a]	-0.05	0.04
Study period	1987–1997	1987–2008	1987–1997	1987–2008
Source	Sun and Zhang (2001)	Mei and Clutter (2010)	Sun and Zhang (2001)	Mei and Clutter (2010)

[a] Significant at the 1% level.
Note: The timberland MLP portfolio in Sun and Zhang (2001) consists of IP Timberlands, Rayonier (including Rayonier Timberlands), Plum Creek, and Pope Resources. The dynamic timberland portfolio in Mei and Clutter (2010) consists of Deltic Timber, The Timber Company, IP Timberlands, Plum Creek, Pope Resources, Potlatch, Rayonier (including Rayonier Timberlands), and Weyerhaeuser. Deltic Timber and Weyerhaeuser were C corporations. Potlatch and Rayonier were publicly traded timberland REITs during only a small portion of the study period. In contrast, our two dynamic timberland REIT portfolios include only firms that were publicly traded timberland MLPs and REITs during our study period.

Table 6.4. Estimated results of the Fama-French three-factor model on timberland REIT portfolios and the NCREIF Timberland Index using quarterly data, 1987–2020

	Timberland REIT portfolio no. 1	Timberland REIT portfolio no. 2	NCREIF Timber-land Index
Alpha (α)	0.019** (0.004)	0.018** (0.005)	0.026** (0.000)
β_{RMRF}	0.797** (0.000)	0.794** (0.000)	0.024 (0.577)
β_{SMB}	0.462** (0.001)	0.480** (0.000)	-0.121 (0.104)
B_{HML}	0.563** (0.000)	0.535** (0.000)	0.028 (0.596)
R^2	0.580	0.589	0.022
R^2 adjusted	0.570	0.580	-0.000
Durbin-Watson statistic	2.130	2.147	1.664
F statistic	60.66	63.03	0.98

** Significant at the 1% level.
Note: Numbers in parentheses are p values.

public-equity timberland returns. The magnitude of β values indicates that the two portfolios are dominated by small to medium firms with high book-to-market ratios. In all cases, the abnormal performances (α values) have dropped slightly, although they are still all significant. These extra factors do not, however, provide additional explanatory power to CAPM in pricing private-equity timberland returns. The only other study that uses the Fama-French three-factor model also reveals that these additional factors provide little additional explanatory power to CAPM in pricing private-equity timberland returns.[19]

While the finding on private-equity timberland investment is similar to what has been found in the literature, the finding of persistent excess return and low systematic risk for public-equity timberland investment warrants explanation. First, unlike in the 2010 study by Bin Mei and Mike Clutter, our two dynamic timberland REIT portfolios consist only of firms that were either timberland MLPs or timberland REITs in the thirty-four-year study period, whereas Mei and Clutter included some C corporations such as Deltic Timber Corporation and Weyerhaeuser Company before 2010. If a firm was neither an MLP nor a REIT during any portion of the study period—such as Deltic Timber Corporation, Weyerhaeuser before 2010, Potlatch before 2006, and Rayonier between 1999 and 2004—it was excluded in our portfolios. Therefore, the timberland REIT portfolios here are pure timberland players.

Second, in the first ten years, the total market value of the timberland REIT portfolios was small because Plum Creek and Pope Resources were small firms then, and only small shares of IP Timberlands LP and Rayonier

Timberlands LP were sold to the public. The latter MLPs were also controlled
by their parent companies.

Similarly, the first timberland REIT did not exist until 1999, and most timberland REITs have existed only since 2004. The returns on our timberland
REIT portfolios thus represent some premiums associated with the conversion of forest products companies from C corporations to S corporations. It is
also possible that TRS contributed to the high returns of the timberland REITs.

Thus, our results for the timberland REIT portfolios and TIMOs show
that with the elimination of the internal subsidy of integrated forest products

**Figure 6.2. Annual real rates of return, the trailing ten-year average of annual real
rates of return, and annualized volatility for the NCREIF Timberland Index and
timberland REITs, 1987–2020**

(a) NCREIF Timberland Index

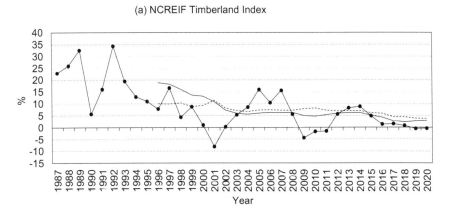

(b) Timberland REITs

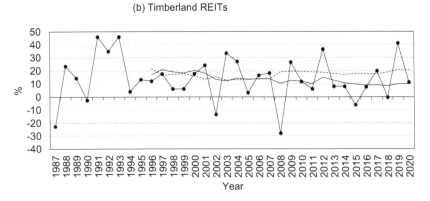

―•―Annual real rate of return (net of inflation) ――Trailing 10-year average rate of return

------ Trailing 10-year standard deviation

companies, improved tax efficiency, and increased concentration on timber-land management, these timberland REITs and TIMOs were less risky and posited better excess returns than forest products companies.[20] On the other hand, the positive α of private-equity timberland returns might be associated with the patience of institutional investors toward embedded strategic options for timberlands and the illiquidity risk they had to bear.[21] The positive α of public-equity timberland returns remains to be explained.

Figure 6.2 presents the real annual rates of return (net of inflation) and the trailing ten-year average of real annual rates of return and annualized volatility for the NCREIF Timberland Index and timberland REITs between 1987 and 2020. It shows that the trailing ten-year averages of NCREIF returns and volatility are stable, although the trailing volatility has slightly surpassed the trailing returns since 2001. On the other hand, timberland REIT returns are more volatile, and the gap between their trailing ten-year volatility and ten-year trailing returns is widening. Again, figure 6.2 confirms that timberland REITs had a higher return and higher volatility than the NCREIF Timberland Index during most of the study period.

In summary, these results show that private equity timberland assets outperformed the market with low systematic risk. Public-equity timberland assets did even better, as they outperformed private timberland investments in terms of returns with elevated risk and were more liquid than private-equity timberland investments.

Timberlands versus Other Financial Assets

In this section I look at what has been revealed by the experience of thirty-four years of institutional timberland investment in terms of the financial characteristics of timberlands—returns, risk, and potential for diversification—compared with other financial assets. After all, timberlands compete with other financial assets for investment.

Because there is overwhelming evidence that most properties in the NCREIF Timberland Index are appraised in the fourth quarter,[22] the index should be measured on an annual basis. Thus, I have used annual data for all financial assets in this section. Furthermore, I have constructed an overall market portfolio that consists of 50 percent common stocks (S&P 500), 10 percent small company stocks (Russell 2000), 15 percent corporate bonds, 15 percent ten-year US government bonds, and 10 percent cash as represented in three-month US Treasury bills. This overall market portfolio is broader than the stock market returns used in the previous section.

Table 6.5. CAPM estimates for the NCREIF Timberland Index, timberland REITs, and other financial assets using the overall market portfolio and annual data, 1987–2020

Asset and period	α		β		R^2	R^2 adjusted
	Coefficient	t-ratio	Coefficient	t-ratio		
Timberland REIT	0.053	1.721*	1.108	4.351**	0.372	0.352
NCREIF Timberland Index	0.079	4.227**	0.058	0.379	0.004	-0.027
Common stocks (S&P 500)	-0.007	-1.222	1.571	35.611**	0.975	0.974
Small stocks (Russell 2000)	-0.021	-1.227	1.599	11.175**	0.796	0.789
NASDAQ stocks	-0.024	-0.614	2.210	9.117**	0.722	0.713
Corporate bonds	0.020	1.725*	0.360	3.779**	0.309	0.287
Ten-year US Treasury	0.016	6.407**	-0.003	-0.163	0.001	-0.030

* Significant at the 10% level.
** Significant at the 1% level.
Note: The overall market portfolio consists of 50% common stocks (S&P 500), 10% small company stocks (Russell 2000), 15% corporate bonds, 15% ten-year US government bonds, and 10% cash as represented in three-month US Treasury bills.

Figure 6.3. The Capital Market Line, 1987–2020

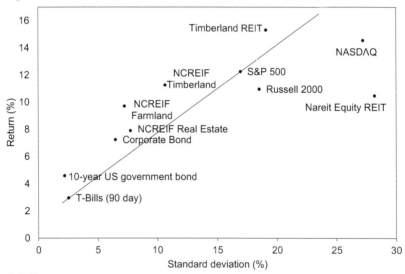

Data Sources:
• Federal Reserve Bank of St. Louis (FRED) for T-Bills (90-day), ten-year Treasury, and Corporate Bond (Merrill Lynch Total Return Index).
• National Council of Real Estate Investment Fiduciaries (NCREIF) for Timberland, Farmland (only from 1991 to 2019), and Commercial Real Estate.
• National Association of Real Estate Investment Trusts (Nareit) for Nareit Real Estate.
• Bloomberg for S&P 500, Russell 2000, and NASDAQ Composite.

Table 6.5 presents the CAPM estimates for private and public timberland equity and other financial assets using the overall market portfolio and annual data between 1987 and 2020. As was true when using quarterly data and stock market returns, I have found that excess returns occurred for both private and public timberland assets and that the systematic risk for both assets was lower than that for all stock indexes but higher than that for long-term US government bonds or corporate bonds.

Figure 6.3 shows the capital market line that compares the level and volatility of the NCREIF Timberland Index and timberland REIT returns to the returns of various financial assets, including real estate. Of these assets, both timberland assets have superior performance—with higher returns and lower volatility—and are clearly above the line that links the annual average return of the risk-free three-month US Treasury bills and that of large US stocks (S&P 500) during the study period. Thus, investing in timberlands in the thirty-four years between 1987 and 2020 generated a risk-adjusted premium.

Another way to look at the return-risk profile of an investment portfolio is the Sharpe ratio. The Sharpe ratio measures the excess return per unit of total risk, as measured by standard deviation. The higher the Sharpe ratio, the better the portfolio performance.

Table 6.6 shows the Sharpe ratio for timberlands, ten-year US Treasury bonds, and the S&P 500 in three separate subperiods and the last thirty-four years. Not surprisingly, the Sharpe ratios for both timberland investment vehicles were higher than that of the S&P 500, and even comparable to that of ten-year US Treasury bonds during the full study period. Thus, both

Table 6.6. Sharpe ratio of ten-year US Treasury yields, S&P 500 total returns, NCREIF Timberland Index returns, and timberland REIT returns

Years	US ten-year Treasury yield	S&P 500 return	Timberland REIT return	NCREIF Timberland Index return
1987–1997	1.72	0.84	0.73	1.72
1998–2008	1.94	-0.01	0.52	0.77
2009–2020	2.74	1.28	0.64	0.67
1987–2020	0.74	0.44	0.66	0.77

Note: The Sharpe ratio or reward-to-variability ratio is a risk-adjusted measure of investment portfolio performance. This measure is defined as (the average returns for a portfolio – the average risk-free rate of return) / the standard deviation of the returns of the same portfolio in an ex post period. It measures the excess return per unit of total risk (standard deviation). The higher the Sharpe ratio, the better the portfolio performance. Thus, both the NCREIF Timberland Index and the timberland REIT outperformed the S&P 500 between 1987 and 2020.

timberland investment vehicles outperformed the S&P 500 during the study period.

The diversification potential of timberland is demonstrated in tables 6.2 and 6.5. Basically, timberland investment—in both TIMOs and timberland REITs—had a significant and positive α and a low β (which were insignificant for the NCREIF Timberland Index) during the study period. The latter means that timberlands and stocks were not correlated and that timberland could be used to diversify an investment portfolio. All these results are consistent with those of previous studies[23] and demonstrate that timberlands differ from other financial assets.

Furthermore, based on the estimates of βs in table 6.5, figure 6.4 presents a security market line between 1987 and 2020 that links the annual return of three-month US Treasury bills and the return of the overall market portfolio defined earlier. Again, it clearly shows that both timberland investment vehicles had higher risk-adjusted returns and did not covary with other financial assets.

Our empirical evidence shows that timberland was a good stand-alone investment, as well as a positive addition to a diversified investment

Figure 6.4. The Security Market Line, 1987–2020

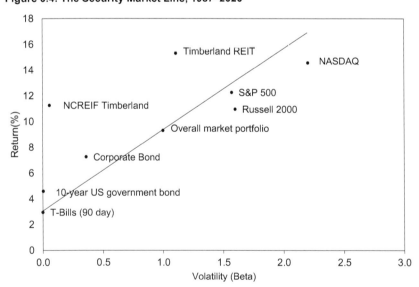

Data sources: see figure 6.3.
Note: The overall market portfolio consists of 50% common stocks (S&P 500), 10% small company stocks (Russell 2000), 15% corporate bonds, 15% ten-year US government bonds, and 10% cash as represented in three-month US Treasury bills. All betas (βs) are from table 6.5.

portfolio. Therefore, the much-discussed benefits of institutional timber-land investment in the 1980s have proven to be true in the thirty-four years between 1987 and 2020. Timberland has established itself as a steady asset class in light of the extraordinary volatility in the broader financial markets during our study period.

TIMOs versus Timberland REITs: Which One Was, Is, and Will Be Better?

The coexistence of both private and public timberland investment vehicles also makes many people wonder which one was, is, and will be better. Based on the empirical evidence above, I would say that the answer depends on investors' main objectives, risk tolerance, and, most important of all, acqui-sition price. Let me start this discussion with retail investors before address-ing institutional investors.

For retail investors interested in timberland investment in the United States, there are only two choices: buy a piece of timberland directly or purchase timberland REIT stocks. Some retail investors may want to be adventurous and experience the fun, contentment, and complexity of being a small private forest landowner. With faith, patience, and frugality, most of them will be as successful as other small private landowners in the long run. This is to say that idiosyncratic risk and uncertainty aside, these small private landowners on average will earn adequate profit and protection against inflation as well as gain satisfaction and enjoyment from nature and entrepreneurship. Needless to say, purchasing the right property at the right location with an appropriate price is critical.

Retail investors who simply want to make money from timberland investment need to be vigilant about the movement of stock prices of the four to five publicly traded timberland REITs in the United States. Timing (and associated pricing) is almost everything in this case. As an example, Weyerhaeuser stock was traded at about $15 per share the first month it became a timberland REIT in January 2010. Within a couple of months, Weyerhaeuser's share price fell more than half before rising drastically to some $20 per share within the next two years, and to $30 per share by mid-2013. The highest price was reached at more than $38 per share in June 2018. Between 2015 and 2020, the price was cut approximately in half from its prior peak three times, before rebounding each time. The stock price of the second largest timberland REIT, Rayonier, performed in a similar fashion.

On average, the price-to-earnings ratio for Weyerhaeuser in the five years prior to August 8, 2020, was 44.7, with a low of 22 and a high of 105 when earnings are measured as trailing twelve months (TTM). The price-to-earnings ratio for Rayonier in the same period was 42.3, with a low of 15.2 and a high of 85. Buying low and selling high are the keys for these retail investors. Having a long-term perspective and a high level of risk tolerance (not overreacting when stock prices fall as well as not buying when stock prices are high) is also important.

Now, for institutional timberland investors who have millions of dollars or more to invest in timberlands, the choices would be investment through TIMOs or publicly traded timberland REITs, or direct investment. Again, in most cases, their assets are managed by professionals in TIMOs or timberland REITs, and I have already presented the features of each vehicle such as liquidity, degree of control, match of interests between management and owners (shareholders), and valuation metrics. Here I just want to reiterate and focus on their recent comparative financial performance.

A valid comparison can be made, especially after 2010 when Weyerhaeuser became a REIT, ensuring similar-sized equity capital invested between timberland REITs and TIMOs in the United States. And more recently, the lines continue to blur as the earnings of Rayonier are now in some small way influenced by its management fees associated with private-equity timberland investment, and REITs such as CatchMark have joint ventures with private-equity timberland investors and are hiring forest management firms to manage some of their properties.

Table 6.7 illustrates just this comparison in the eleven years after Weyerhaeuser became a REIT. The results show that timberland investment returns lagged the broad markets as measured by the S&P 500 in the last eleven years. As far as the two timberland investment vehicles are concerned, the results are very similar to those of the last twenty years: public timberland equity had a much higher return and was much more volatile than private timberland equity. Again, this is related to the differences in construct and leverage in timberland REITs and the NCREIF Timberland Index. Properties in the NCREIF Timberland Index are appraised by artificial construct. Further, the timberland REITs generally have a debt-to-total-asset value in the range of 30–50 percent or higher. This would increase the total returns on equity as well as its level of volatility.

If we know about the differences in construct and leverage, the simple answer to our question—which one was better in the last ten to eleven

Table 6.7. Annual rates of return for ten-year US Treasury, S&P 500, NCREIF Timberland Index, and timberland REITs, 2010–2020

Year	Ten-year US Treasury yield	S&P 500 return	NCREIF Timberland return	Timberland REIT return
2010	3.30	15.06	-0.15	13.42
2011	1.89	2.11	1.57	9.30
2012	1.78	16.00	7.75	38.50
2013	3.05	32.39	9.69	9.45
2014	2.17	13.69	10.48	9.59
2015	2.27	1.38	4.97	-6.07
2016	2.45	11.96	2.59	8.62
2017	2.40	21.83	3.63	21.08
2018	2.69	-4.38	3.21	-32.50
2019	1.92	31.49	1.30	40.00
2020	0.89	18.40	0.81	12.44
Average	2.25	14.54	4.51	11.67
Standard deviation	0.66	11.70	3.65	20.98
Coefficient of variation	0.29	0.80	0.81	1.80
Average for 2016–2020	2.07	15.86	2.75	8.15

Data sources: see table 4.2 and text.

years—would still depend on investors' risk-return trade-off metrics. What about now and in the next few years or decade? This is to say, if institutional investors have money and want to invest in timberland now or soon, which vehicle should they choose? The answer to this question would require some sophisticated analyses. There are two ways to compare and thus make a judgment or educated guess on these two vehicles.

The first is to compare the immediate past, say the last five years. Even though past performance does not guarantee future returns, the immediate past performance may still give investors some guidance in the short term because of momentum or inertia.

As noted earlier, Weyerhaeuser and Rayonier stocks respectively traded at average price-to-earnings ratios of 44.7 and 42.3 in the five years prior to August 8, 2020. Neither one was cheap, because the implied average return on equity for these two stocks was only about 2.23 and 2.36 percent, respectively. Nonetheless, the dividend yields for these two firms (before Weyerhaeuser announced suspension of dividends on May 1, 2020) were about 4.5 and 3.8 percent, respectively, during the same period. This would round the total returns for investing in these two firms to 6–7 percent, which is higher than the returns of the NCREIF Timberland Index in the five years prior to August 8, 2020.

Interestingly, as table 6.7 shows, despite huge volatility, the average rate of return for our timberland REITs was 8.15 percent in the five years between 2016 and 2020. One could say this was not bad for a low interest rate environment. In comparison, the returns on timberland REITs more than doubled the rate of return of the NCREIF Timberland Index (2.75) in the same five years. But both timberland investment vehicles lagged in their performance compared to the S&P 500 by a big margin during this period.

Second, one could do a valuation exercise to estimate the per-acre timberland value of one vehicle compared to another. Richard Hall, in his capacity as instructor of a forest finance course at Auburn University, did just that in the summer of 2020. I use his example here as an illustration to compare the per-acre value of US timberland held by TIMOs and several timberland REITs and to possibly identify which one is relatively undervalued.

Because NCREIF reports the per-acre US timberland value on average and by region, one needs to estimate only the per-acre value of timberland owned by REITs. Thus, the valuation exercise starts with identifying the respective enterprise value of the four US timberland REITs. Enterprise value is defined as market capitalization in equity plus outstanding debt.

Next, one needs to estimate the non-US timberland value, which includes both timberlands owned in other countries and nontimberland assets in the United States, such as TRS, for each of these REITs. Sound professional judgment needs to be applied in this step. The difference between the enterprise value and non-US timberland value is the total value of US timberlands owned by a timberland REIT. Finally, dividing the total value of US timberland by the total US timberland acreage gives us the per-acre value of US timberland owned by each timberland REIT, as shown in table 6.8.

Table 6.8 illustrates that timberlands held by REITs were about fairly valued compared to those held in private equity as of December 31, 2019. However, they might be slightly undervalued if one considers that timberlands held by the former had more mature trees than those held by the latter. Furthermore, real estate appraisals have historically lagged markets both up and down. It is likely that this also occurs for timberland appraisals.

All these reasons may lead some investors to believe that the cheaper place to buy timberland in the United States is on Wall Street since timberland

Table 6.8. Timberland valuation: Timberland REITs versus the NCREIF Timberland Index

Company	Enterprise value ($ billions)	Non-US timberland value ($ billions)	US timberland acres (millions)	Total value of US timberland ($ billions)	Value of US timberland ($/acre)
Public timberland (as of December 31, 2019)					
Weyerhaeuser	28.90	3.50	11.00	25.40	2,309.28
Rayonier	5.42	1.00	2.18	4.42	2,028.91
PotlatchDeltic	3.60	0.90	1.83	2.70	1,472.14
CatchMark	1.01	0.25	0.44	0.76	1,735.40
NCREIF – Private timberland (as of December 31, 2019)					
US	23.17	0.00	12.36	23.17	1,873.81
Southeast	14.87	0.00	8.22	14.87	1,810.34
Pacific Northwest	6.64	0.00	2.33	6.64	2,853.42
Northeast	0.92	0.00	0.74	0.92	1,241.32
Lake States	0.62	0.00	0.97	0.62	635.03
Public timberland (as of March 19, 2020)					
Weyerhaeuser	17.98	3.50	11.00	14.48	1,315.96
Rayonier	3.86	1.00	2.18	2.86	1,313.45
PotlatchDeltic	2.57	0.90	1.83	1.67	911.34
CatchMark	0.72	0.25	0.44	0.47	1,080.60

Source: Richard Hall, personal communication with author, March 21, 2020. Hall drew data from Bloomberg, individual timberland companies' annual reports and websites, and 2019 NCREIF timberland reports. Non-US timberland assets include manufacturing and distribution facilities, long-term forest leases in the US or elsewhere, foreign timberland, and other assets that are not representative of full ownership of US timberland assets. Partial ownership of US timberland assets is considered and addressed as part of the final calculation. Valuation approaches for non-US timberland assets can be quite variable.

REITs might be valued at a discount compared to private equity valuations based on third-party appraisals, as of December 31, 2019. Of course, just as lags and imprecision could occur in private timber appraisals, our timberland valuation results could be subject to challenges on the part of non-US timberland assets that include both timberlands overseas and assets in TRS. And there is no way to know the actual timber inventory for timberlands held by TIMOs or even REITs.

Fast-forward to March 19, 2020, when most US stocks were trading near their bottom after a sharp decline in the wake of the COVID-19 global pandemic, and the timberlands held by publicly traded REITs were undervalued by more than 35 percent compared to privately held timberlands in the

NCREIF Timberland Index (table 6.8). It was an extraordinary opportunity to buy publicly traded timberland REIT and other stocks. Those who were bold enough to purchase timberland REIT and other US stocks were greatly rewarded a few months later.

The lessons? Timing is everything. Being aggressive when others are scared is also critical, especially if one is considering investing in timberland REITs.

CHAPTER 7

Institutional Timberland Investment around the World

In this chapter, I present the history and development of global timberland investment by institutions in the last thirty-plus years. Global timberland investment is important in two aspects. First, as *New York Times* columnist Thomas Freeman says, the world is flattening. Certainly, the capital and commodity (including forest products) markets are more globally integrated today than ever before. If timberland investment in one country has high risk-adjusted returns, investors from other countries can invest there, just as the British Coal Board and Fountain Forestry did in the United States in the late 1970s and early 1980s, and as some US investors did in New Zealand, Chile, and Australia in the early 1990s. The allocation of an investment portfolio has gone global for some institutional investors.

Second, in today's globalized economy, forests in one part of the world are linked to forests elsewhere through expansion of international trade, investment, enhanced communication, and global environmental changes. Events associated with forests in one place could have an impact on global forestry sustainability. A good example is the listing of the northern spotted owl as a threatened species in the United States in 1990, which triggered a substantive reduction in the timber harvested from public lands in the US Pacific Northwest.[1] While the event caused US timber prices to rise and stay high for about seventeen to eighteen years and thereby enhanced timberland investment returns in the United States, it also increased imports of lumber and other forest products from Canada and elsewhere into the United States, raising the sustainability question in Canadian forests.[2] More recently, a logging ban by the Chinese government on its natural forests spiked investment in planted forests in the country by domestic and international sources and led China to import up to 40 percent of its roundwood needs for domestic consumption and reexport. Without doubt, international timberland investment has further strengthened such linkages in global forest sustainability.

Broadly, the theory guiding institutional investors investing in timberlands in other countries is the same as that guiding domestic investors: seek higher risk-adjusted returns. Using US investors investing in other countries to seek higher risk-adjusted returns in US dollars as an example, this thesis is based on three considerations.

The first is the relatively higher timber market inefficiency and capital shortages in other countries that offset the associated country and exchange rate risk. Such an investment is complicated for forestry investors because the asset base is necessarily anchored in the foreign country, which may limit capital outflow. This leaves potential US investment destinations only to countries that either are import substitution candidates such as China or have natural export advantages such as New Zealand. Although investors seek gains from market inefficiency, the normal functionality of timber and timberland markets is critical, as some countries may have too many regulations and red tape surrounding timber management activities, raising the transaction costs of doing forestry business there. Furthermore, the market inefficiency argument may not equal higher forest growth rates, which often lead to higher land prices. On the other hand, if a country has a higher growth rate in demand relative to supply, it is a good reason for it to attract international monies, all else being equal.

The second is added diversification, which may or may not be the case. This depends on market dynamics and the functional currency of revenues and costs. Costs are generally small and denominated in local currency with the exception of fuel, which is in US dollars in many countries. Revenues associated with various forest products (sawtimber, pulpwood) may or may not be in US dollars. This is to say that the added diversification of investing in timberlands in other countries is also complicated by the added layer of exchange rate conversion.

Third, there is the question of possible reputational risk. As we will see later in this chapter, there were protests against a few US and European institutional investors regarding their investment activities in Latin America and Africa.

The next section starts with foreign timberland investment in the United States, paying special attention to the Foreign Investment in Real Property Tax Act of 1980 (FIRPTA), followed by global investment by US-based TIMOs and REITs. The third section covers non-US-based TIMOs and their timberland investment on behalf of institutional investors outside the United States, notably in Europe, Latin America, Africa, and Asia. The fourth section presents publicly traded timberland and forestry funds. The final section discusses the implications of international timberland investment for global forest sustainability.

Foreign Institutional Timberland Investment in the United States

In addition to British Coal and Fountain Forestry, sovereign funds also invested in US timberlands in the 1970s and 1980s. For example, the government of Kuwait bought some timberlands in the US South. China, through a state-run company—the China International Trust and Investment Corporation (CITIC)—bought some timberlands in the US Pacific Northwest, facilitated by The Campbell Group. Most of these investments did not last long. The Kuwait government sold its timberland assets to rebuild the country in 1992 after the invasion by Saddam Hussein and the Gulf War in 1991. CITIC also sold its timberland holdings in the 1990s and moved its monies to New Zealand and elsewhere.

Since 1990, most foreign timberland investment in the United States has occurred through US- or foreign-based TIMOs, or by purchase of shares of publicly traded timberland REITs. For example, International Woodland Company (IWC), a Denmark-based forestry consultant and adviser, has used several US TIMOs to place investment commitments it received from European clients. In 2013, BTG Pactual, a Brazilian investment management firm, bought RMK Timberlands from Regions Bank and set up BTG Pactual Timberland Investment Group. By 2020, BTG was one of the largest TIMOs, with more than $3.5 billion in assets under management (table 5.1).

Similar to US institutions, non-US institutions investing in US timberlands look for adequate returns, low risk, and the potential for diversification. However, they face an added layer of tax that US institutions do not have—they must pay foreign real property tax when they dispose of their US real property interests (USRPI) under FIRPTA. FIRPTA imposes income tax on foreign persons disposing of USRPI, which include direct timberland ownerships and stocks in timberland REITs.

All else being equal, foreign investments in US timberlands might have sputtered after the enactment of FIRPTA for two reasons. First, FIRPTA imposes US federal tax and return filing obligations on non-US investors with respect to the disposition of USRPI. FIRPTA treats gains from the disposition of USRPI by foreign persons or institutions as if they were engaged in a US trade or business with "effectively connected" gains that are subject to US tax. This tax thus affects the net returns of foreign investments in US real estate, including timberlands.

Second, FIRPTA creates uncertainty with respect to timberland REITs, especially private timberland REITs. Section 897(h) of the IRS Code applies

a "look-through" rule to distributions by a REIT from gains from dispositions of USRPI. The question is whether ongoing gains from dispositions
of standing timber are gains from a disposition of USRPI in the first place.
Conventional wisdom says "yes." But the purchaser in a retained economic
interest contract never actually owns the standing timber, and the risk of
loss and benefit from timber growth remains with the seller until the timber is cut. Revenue Ruling 2001-50 treats the standing timber as separate
from the underlying land. However, FIRPTA as applied to dispositions of
standing timber predates, and thus does not take into account, this ruling.
Also, the IRS Code apparently requires withholding on these distributions
at 35 percent instead of the 15 percent (or 20 percent after 2010) capital
gains rate. All these regulations and uncertainty could have prevented for-
eigners from investing in timberlands in the United States on a larger scale
than otherwise.[3]

FIRPTA tax is imposed at regular tax rates for the type of taxpayer on the
amount of gains recognized. Purchasers of real property interests previously
held by a foreign entity are required to withhold tax on payment for the property. The withholding of the gross sale proceeds payable to non-US investors
before 2016 was 10 percent, which was changed to 15 percent under the
Protecting Americans from Tax Hikes Act of 2015.

On the other hand, FIRPTA excludes from US tax any income earned by
foreign governments, including sovereign worth funds and certain non-US
pension plans formed for the benefit of foreign government employees. In
particular, the Protecting Americans from Tax Hikes Act of 2015 exempts
from FIRPTA dispositions of USRPI held directly or indirectly by "qualified foreign pension funds" and their wholly owned subsidiaries. A non-US
pension fund is "qualified" if it meets a number of technical requirements,
including that it be subject to regulation and reporting requirements about
its beneficiaries in its jurisdiction, have no beneficiary with a right to more
than 5 percent of its income or assets, and benefit from tax breaks on contributions or investments in its jurisdiction. Such a pension fund will, however,
continue to be subject to the general rules imposing US federal tax and return
filing obligations with respect to US trade or business activities.

Also, the Protecting Americans from Tax Hikes Act of 2015 increased
from 5 to 10 the percentage of publicly traded stock that a non-US investor
may hold in a public REIT (but not any other type of US real property holding
company) and still qualify for an exemption from FIRPTA on sale of stock
of, and distributions by, the public REIT. It also provided an exemption from

FIRPTA for certain treaty-eligible publicly traded entities that invest in REITs even if the 10 percent limitation is exceeded.[4]

Finally, US treaties with Canada, Germany, Mexico, the Netherlands, and the United Kingdom reduce the withholding rate to zero if investors residing in these countries are tax exempt. The sale-lease-back arrangements between Travelers and Bowater also serve as a way for foreign investors to avoid FIRPTA.

US-Based Institutions Investing in Timberlands Elsewhere

US-based forest products companies had owned timberlands in many other countries where they had forest products manufacturing facilities long before institutional investors started to look for timberlands there. The initial phase of international timberland ownership by US-based forest products companies—from the late 1890s until the Second World War—occurred mostly in Canada to supply timber and forest products to US markets. For example, International Paper Company started to own forestlands and lease Crown lands in Canada in 1913. Other US forest products companies followed suit in Canada, and in 1990, US companies owned more than 30 percent of paper company assets in the country.[5]

A second phase occurred between the end of the Second World War and 2000, when US-based forest products companies invested in South America, Europe, Oceania, and Asia as they expanded their businesses there.[6] In the early 1990s, when the shortage of timber occurred in the US Pacific Northwest, some US forest products companies bought more timberlands in the southern United States but also in countries in the Southern Hemisphere, including Brazil, Chile, New Zealand, South Africa, Uruguay, and Argentina. Included in this phase of international industrial timberland ownership expansion was investment by large US companies, including International Paper, Weyerhaeuser, Champion International, Stone Container, and the privately owned Simpson Paper. The motivation for these companies to invest in timberlands overseas was to meet the demand for timber by their mills in these countries, some of which were newly established as solely owned or as a joint venture with local forest products firms.

As US forest products companies began to sell more of their timberlands in the United States in the late 1990s, they also started to exit from timberland ownership in Europe, Oceania, and South America. For instance, International Paper sold all of its 820,000 acres of timberland in New Zealand in 2005, and Weyerhaeuser did the same in 2008. As in the United States, the buyers of overseas assets were mostly US institutions through TIMOs. In the case of Rayonier, its timberlands in New Zealand were acquired when it was an MLP in 1992. DANA Limited, a New Zealand–based consulting and investment firm that tracks international timberland trends, identified US-based TIMO ownership of 1.1 million acres in New Zealand and 1.6 million acres in nine other countries in 2005.[7] The same firm estimated that US-based TIMOs owned around 1 million acres, or 39 percent, of "industrial" scale forest estate in New Zealand in 2016, down from 48 percent in 2013. On the other hand, investors from other countries, notably Australia, Canada, China, Japan, Korea, Malaysia, Switzerland, and the United Kingdom, owned some 37 percent of the industrial forest estate, and the remaining 24 percent belonged to New Zealanders in 2016.[8]

The initial attraction of investing overseas for US institutional investors was the prospect of higher returns in countries associated with fast-growing plantations, which offset country risk and thus spread the risk of their total investment portfolios. Consequently, countries in the Southern Hemisphere with a relatively stable political system and a good investment climate—such as New Zealand, Australia, Brazil, Chile, and Uruguay—were the earliest destinations of US institutional monies. As in some parts of the United States, there was inefficiency in timber and timberland markets in these countries in the 1990s and early 2000s. Later, US-based and other TIMOs looked at emerging markets where the supply and demand imbalance in timber was more promising and where greater market inefficiencies existed, such as in Africa, Central America, and eastern and central Europe.

* * *

The first overseas transaction by a US-based TIMO was in New Zealand in 1992 by what was then Resource Investments Inc. (RII, now Global Forest Partners). The investment was triggered by the decision of the New Zealand government to privatize its planted forest assets, which accounted for about half of the 1.2 million hectares of planted forests in 1989. Most of the planted forests in New Zealand are radiata pine, which originated on the Monterey Peninsula in California in the mid-1800s. Between 1990 and 1992, the New Zealand government sold more than 350,000 hectares of planted forests to the private sector, and another 188,000 hectares in 1996.[9] This privatization was triggered by the Labour government, which found itself in a stark financial situation after winning the general election in 1984. The government first started to corporatize government departments and then to privatize the "exotic" (radiata) forest estate it owned. But a big impediment was Maori land claims. Thus, it created "Crown Forest Licenses" and sold off what were essentially long-term timber deeds, mostly for ninety-nine years.

The largest private forest products company in New Zealand at the time was Fletcher Challenge, which had forestry, forest products manufacturing, building materials, and petroleum operations in New Zealand and elsewhere. Fletcher Challenge had owned forest plantations and bought a lot of wood from government forest plantations before the privatization of public forests started in 1989. When the New Zealand government started to sell its forests, Fletcher Challenge wanted to purchase a particular area of plantation—some 50,000 hectares near the city of Nelson in the northwest part of South Island in 1990. Fletcher Challenge already owned nearly 10,000 hectares of planted forests in the area; getting these additional Crown forests would increase the scale of its operation there and boost its future revenues and profit.

The problem was that Fletcher Challenge did not have all the financial capital it needed, which was about NZ$225 million (or US$125 million when the deal was closed in the summer of 1992). Fletcher Challenge wanted a joint venture with an investor on a fifty-fifty basis. So it asked Goldman Sachs to identify potential coinvestors in the United States. Two individuals were instrumental in this process—Dennis Neilson, the director of marketing, and Bryce Heard, then CEO of Fletcher Challenge. Goldman Sachs brought RII to Fletcher Challenge. TIMOs were just coming out of their infancy, and only RII was willing to go overseas. To sweeten the deal, Fletcher Challenge offered its own forests of about 10,000 hectares on top of the 50,000 hectares of government lands, making the whole package 60,000 hectares, or nearly 150,000 acres.

After being convinced, RII promised to get the other half of the money needed within a year. Based on this promise, Fletcher Challenge started to bid and eventually won the bid for NZ$225 million in 1991. In the meantime, Dennis Neilson and the chief financial officer of Fletcher Challenge, along with Ed Broom and Eric Oddleifson from RII and one person from Goldman Sachs, went on to tour around the United States and try to raise the money from institutional investors. They had to sell the concepts of New Zealand, forestry, radiata pine (and a little Douglas fir), and Fletcher Challenge together, mostly to pension managers and consultants.

In the end, RII received support and commitments from five or six pension funds, and the joint venture was valued at NZ$250 million.[10] This meant that the 10,000 hectares of fee simple land that Fletcher Challenge owned were valued at only NZ$25 million, which was only about 10 percent of the total asset value. Perhaps the planted forests that Fetcher Challenge owned were a bit younger than the Crown forests. On the other hand, they were on fee simple land, whereas the forests under Crown Forest Licenses were a long-term lease. Certainly, RII bargained hard, but Fletcher Challenge could not, because it wanted the capital investment to pay down debt incurred in its NZ$1.3 billion "disastrous" purchase of UK Paper PCL (UK Paper) in the United Kingdom in 1990.[11]

It was a year and a half before both sides closed the deal in June 1992. The time was spent fund-raising, drafting legal paperwork, and especially answering the question of how to properly do the valuation if one partner wanted to get out of the joint venture later. As for the latter, they came up with an ingenious solution—one party would divide the assets into two halves and the other party would have the right to first pick.

This joint venture, called the Nelson Forests Joint Venture, was very successful in its early stage, as the annual rate of return between 1992 and 1996 was 11.4 percent. In June 1994, RII made a second investment in New Zealand. This time, it partnered with Rayonier and bought over 250,000 acres of planted forests there in 1992. The deal was again a fifty-fifty joint venture.[12] Rayonier, now a REIT, still owns some of the original timberlands it bought back in 1992.

The connection between RII and Fletcher Challenge also led RII to become the first US-based TIMO to invest in Chile through another joint venture with Fletcher Challenge, called Forestal Bio Bio, in 1993. The main assets of Forestal Bio Bio were 40,000 hectares of timberland and a small paper mill in Chile.

* * *

Not all timberland investments in New Zealand in the 1990s were successful. In the mid-1990s, Xylem Global Partners LLC (Xylem), based in New York, had two timber funds in New Zealand and several other countries and made numerous investments, including shareholdings in Fletcher Challenge and Evergreen Forest Products Company of New Zealand. Both timber funds performed poorly, and as a result, Xylem was fired by its investors in 2003, who then hired GFP to liquidate its portfolio. Xylem eventually went bankrupt during the 2008–2009 financial crisis. Originally a TIMO, by the time of its demise, Xylem was neither a TIMO nor a REIT. Rather, it was a hedge fund that used an arbitrage model and actively took long and short positions of public forest products equity.

A joint venture between Fletcher Challenge and CITICFOR Inc. (a subsidiary of CITIC of China), called the North Central Island Forest Partnership (NCIFP), was created to purchase the largest sale of assets during the privatization of Crown forests in 1996. While it was structured similarly to the joint venture between Fletcher Challenge and RII in 1992, this later venture lasted only a little more than three years.

When first created, the consortium of investors consisted of Fletcher Challenge (through its forestry division, Fletcher Challenge Forests Ltd., which managed the business, 37.5 percent), CITICFOR Inc. (37.5 percent), and a UK firm called Brierley Investments Ltd. (25 percent). In 1998, Brierley chose to exit, as the joint venture agreement gave it the right to leave after two years, forcing Fletcher Challenge and CITICFOR to buy its 25 percent share. Thus, Fletcher Challenge and CITICFOR each owned 50 percent afterward.

The partnership (NCIFP) covered 188,000 hectares of planted forests (including the Kaingaroa State Forest) in the central region of North Island as well as processing plants, a nursery, and a seed orchard in 1996. The partnership eventually paid NZ$2.026 billion for these assets, which was 25 percent higher than the next highest bid.

The Kaingaroa State Forest was adjacent to Fletcher Challenge's own 117,000 hectares (289,000 acres) of plantation in central North Island and was first put out for sale by the New Zealand government in 1990–1991. That sale failed because there was no taker. Since Fletcher Challenge bought UK Paper in 1990–1991, it did not have the money to bid. Had Fletcher Challenge bought the Kaingaroa State Forest instead of UK Paper, it would probably still have been standing instead of being split into small pieces in 2001.

In 1996, the 188,000-hectare planted forest assets alone were valued at NZ$1.8 billion or NZ$9,574 per hectare. In contrast, the average price of the 246,706 hectares sold by the New Zealand government between 1990 and 1992 was only NZ$4,053 per hectare. This is an indicator of inflation, and perhaps timber growth, as well as overbidding on the part of the whole consortium (NCIFP). This overbidding was a critical mistake. Combined with debt financing that represented 60 percent leverage and mismanagement, it meant there was no chance for the joint venture to succeed given the subsequent decline in timber prices (fig. 7.1).

Interestingly, Fletcher Challenge initially submitted a bid on its own for NZ$1.2 billion and was only the third highest bidder. While the highest bidder offered NZ$1.6 billion, the offer was slightly short of the NZ$1.8 billion assessed value, and thus the initial sale did not go through. How Fletcher

Figure 7.1. Average quarterly pruned and unpruned radiata pine log prices for export in New Zealand, 1992–2020

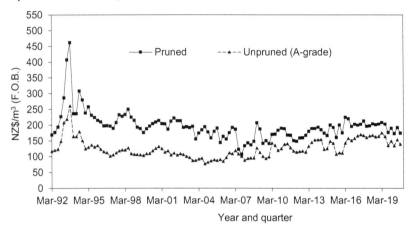

Source: Forestry New Zealand, accessed January 23, 2021, https://www.teururakau.govt.nz/news-and-resources/open-data-and-forecasting/forestry/wood-product-markets/.

Challenge later got the property with two other partners with an offer of more than NZ$2 billion is of interest to this author and perhaps many readers of this book.

One could imagine that there must have been some negotiations between the New Zealand government, which was under public pressure to sell the assets to a domestic firm, and Fletcher Challenge, which then brought two additional partners and increased its bid price based on some hint or inside information. The New Zealand government certainly wanted to get the highest price for its assets. But why the consortium—especially Fletcher Challenge and CITICFOR because they had higher shares and had to take the third partner's share if needed in two years—agreed to pay 11 percent more than the assessed value, 25 percent more than the next highest bid, and 69 percent more than Fletcher Challenge's original bid in the previous round has been a puzzle to many investors and observers.

Things then got worse. In July 1998 the partnership announced it was negotiating to restructure over NZ$1.2 billion in debt after timber (especially Douglas fir) prices fell. The partnership's cash flow forecasts were based on heavy early harvesting of Douglas fir. Unfortunately, harvesting had to be reduced when Douglas fir log prices collapsed and were followed by a collapse of radiata pine log prices.

By 2000, NCIFP had been dissolved acrimoniously and receivers were appointed the following year, after it failed to make payments on its NZ$1.48 billion debt. The receivers went looking for buyers.

Falling log prices, helped by rising New Zealand dollars, put pressure on the NCIFP receivers. As an indication, in July 2003, Carter Holt Harvey, another New Zealand forest products company, announced that it would write down the value of its forests by 31 percent. Fletcher Challenge had valued the NCIFP forests at NZ$1.5 billion. The asking price of NCIFP was reported at just above NZ$1 billion (US$650 million), which was the amount the NCIFP owed to a syndicate of banks.

The sale formally took effect in December 2003, for a price of $NZ1.081 billion. The purchase was under the management of GMO Renewable Resources on behalf of the Harvard Endowment (HMC), which made a fortune on it.[13]

This story shows how overbidding, excessive debt financing, mismanagement, and the near-term decline in timber markets could bring down a business venture that had valuable forest assets. In particular, no financing or management could overcome overvaluation and overbidding for a timberland property and heavy debt financing. One could argue that if all the other factors had unfolded as they did but there was no debt, the property would still be owned by the consortium. It also shows what clever investors with money like HMC can do when an undervalued opportunity arises.

* * *

Building on its joint venture experiences with Fletcher Challenge in New Zealand and Chile and with Rayonier in New Zealand, RII (now UBS RII) teamed up with Weyerhaeuser to make global timberland investments in 1995.[14] Weyerhaeuser and UBS RII planned a fifty-fifty joint venture partnership to expand timberland ownership overseas, and UBS RII set off to raise its share of the capital. On May 1, 1997, UBS RII announced in a news release that the partnership had raised $524 million, of which over $260 million came from thirty-two institutional investors.[15] Peter C. Mertz, who had previously worked at Weyerhaeuser and International Paper and then served as managing director and chief investment officer at UBS RII, was instrumental in creating this partnership.

In the news release, UBS RII stated that the $524 million joint venture partnership with Weyerhaeuser Company "will make investments in timberlands and related assets outside the U.S. The primary investment focus will

Figure 7.2. Average forest growth rates among countries in the 1990s

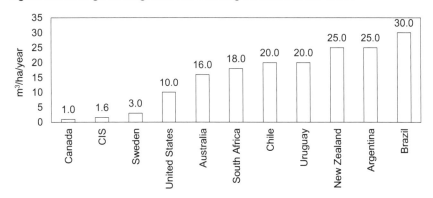

Source: UBS RII (1996a).
Note: "CIS" stands for Commonwealth of Independent States (after the breakup of the former Soviet Union).

be in planted, softwood forests in the Southern Hemisphere." The investment strategy was to invest in fast-growing, investment-grade, primarily softwood plantations. Targeted countries included Argentina, Australia, Brazil, Chile, New Zealand, and Uruguay, where planted forests grew two to four times as fast as in the United States (fig. 7.2); where the labor, land, and energy costs were low; and where investment opportunities were many. Through this joint venture, known as the RII Weyerhaeuser World Timberfund, UBS RII was able to access Weyerhaeuser's expertise in silviculture, processing, marketing, and distribution. UBS RII hoped that combining all these factors would create the potential to produce a 12–14 percent real rate of return annually (fig. 7.3) and achieve portfolio diversification in region, species, age class, and end market.

In addition to this partnership with Weyerhaeuser, UBS RII agreed in 1997 to Weyerhaeuser's buyout of Fletcher Challenge's interest in the original New Zealand joint venture in 1992. The two partners in the Nelson Forests Joint Venture later acquired and substantially upgraded a sawmill. A year earlier, RII Weyerhaeuser World Global Timberfund had also made a bid for the aforementioned Kaingaroa State Forest in New Zealand but was not successful. Through the World Timberfund, UBS RII and Weyerhaeuser were also the first group of US investors to purchase land and establish softwood plantations in Uruguay beginning in 1997. And in 1999, they successfully acquired Green Triangle Forest Products Ltd., which owned over twenty thousand hectares of planted radiata pine timberland and a pine sawmill in Australia.

Figure 7.3. Country risk-return profile: RII's return requirements from biological growth

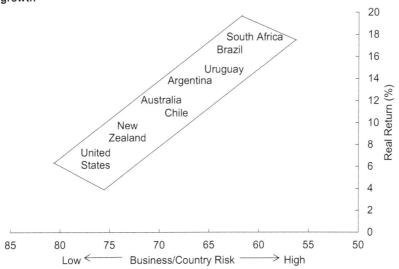

Source: UBS RII (1996a). The x-axis represents a country's business-risk index; high numbers represent low levels of business risk.

By 2006, GFP and Weyerhaeuser began unwinding their joint ventures, beginning with Nelson Forests. After the conclusion of a sales process that failed to provide a value satisfactory to both partners, GFP offered to buy out Weyerhaeuser the following year. Nelson Forests was recapitalized by GFP with new investors alongside those wishing to remain. Nelson Forests continued under the sole ownership of institutional investors until 2018, when it was sold to OneFortyOne Plantations Holdings Pty Limited (OFO). OFO is an Australian company formed by institutional investors when they acquired Forestry SA during its privatization by the state of South Australia in 2012. OFO is advised by Campbell Global. Thus, Nelson Forests has been owned by institutional investors for nearly thirty years.

GFP and Weyerhaeuser also dissolved their other joint ventures, including the World Timberfund, through a series of sales and partitioning of assets in Australia and Uruguay. In Australia, they sold the sawmill within the joint venture to Carter Holt Harvey, the New Zealand–based forest products company (which was then owned by private equity investor Graeme Hart), in 2008, following a buyout by GFP of Weyerhaeuser's interest in the forests and recapitalization similar to that of Nelson Forests the prior year. Today that sawmill is owned by OFO.

In Uruguay, the partners partitioned their forest holdings and continued to operate their businesses separately. In 2017, Weyerhaeuser sold its forests and plywood mill in Uruguay to institutional investors advised by BTG Pactual Timber Investment Group.

* * *

Also around 1993, Hancock Timber Resource Group acquired timberlands in British Columbia, Canada. In 1996, Hancock expanded to Australia, and to New Zealand in 2004. PruTimber and GFP both made investments in Brazil around 2002. In the early 2000s, The Campbell Group made investments in Australia. Resource Management Service and GreenWood Resources made timberland investments in China in the mid-2000s.

As noted earlier, GMO Renewable Resources also invested in New Zealand on behalf of the Harvard Endowment in the early 2000s. But New Zealand was a relatively small market, and the fall of radiata pine log prices in addition to other circumstances in the early 2000s tempered enthusiasm for investing there. Australia then became popular in the late 1990s and early 2000s. Hancock was the earliest and biggest US-based owner of timberlands there, and GMO Renewable Resources (now The Rohatyn Group) was probably the second or third largest at one time. Uruguay became popular in the early 2000s as well, as European and North American companies and TIMOs (including GMO Renewable Resources, GFP, and RMK Timberland Group) and others, including the Harvard Endowment, started to invest there. This caused the land value in Uruguay to escalate rapidly, with plantation land prices several times higher in 2006 than even just five years earlier. Many of these new timberland investments were made at that time, aiming to supply timber for two new world-class pulp mills built by UPM and Stora Enso (in a joint venture with the Chilean company Arauco), which became operational in 2007 and 2014, respectively. In July 2019, UPM announced that it would invest US$3 billion in a new pulp mill in Uruguay, scheduled to be completed in 2022.[16]

In 2006, Hancock bought 250,000 hectares of timberland from Carter Holt Harvey in New Zealand in an auction. The forests had a book value of around NZ$1.5 billion and were located in Northland, Auckland, central North Island, Hawke's Bay, and Nelson. About 100,000 hectares of the land was freehold, and the rest was nonfreehold land, standing timber, plants and equipment, business contracts, licenses, and contents.[17]

In 2012, Campbell Global led a consortium to purchase over 250,000 acres of forestry assets in the Green Triangle plantations of South Australia.[18]

The purchase entitled the consortium to manage and generate income from the assets for three rotations, or approximately 105 years. In the same year, Campbell Global set up a perpetual fund with Australian institutional investors and their portfolio company (OneFortyOne) to manage these assets. In 2018, OneFortyOne purchased Nelson Forests in New Zealand from GFP, as noted earlier.[19]

US-based institutional investors also made timberland investments in central and eastern Europe as countries there joined the European Union, as well as in Africa.[20]

By 2010, strong competition for timberland opportunities in traditional global investment geographies or the so-called mature markets was eliminating the low-hanging fruit there. Consequently, US-based TIMOs went farther afield to emerging markets, such as Cambodia (GFP), Panama (The Forestland Group), and central and eastern Europe (Harvard Management Company, GreenWood Resources). Thus, US-based TIMOs went to Oceania and South America first in their investment overseas. Gradually they expanded to Central America, Asia, central and eastern Europe, and Africa.

The largest US-based TIMO investing in Africa used to be the Global Environment Fund (GEF), which should not be confused with Global Environment Facility, which has the same acronym and was created at the 1992 Rio Earth Summit to help tackle the most pressing environmental problems on our planet. Established in 1990, the Global Environment Fund is a global alternative asset manager and has grown into one of the world's most successful investment firms dedicated to the energy, environmental, and natural resources sectors. Unlike typical TIMOs that raise funds from pension funds, endowments, and wealthy individuals, GEF raised money mostly from the International Finance Corporation and other multilateral development finance agencies. Therefore, most of its investments have explicit or underlying sustainability and/or social development goals, in addition to financial returns.

GEF's first general emerging market fund for investing in timberland and sawmill assets in Africa was in South Africa in 2001. Then it started a second fund and went to Malaysia and Argentina in 2007, Chile in 2008, Mozambique in 2009, Tanzania and Ghana in 2011, Swaziland in 2012, Uganda in 2015, and Gabon in 2016. In 2016, GEF developed an African timberland investment strategy and set up an Africa-only fund—the Africa Sustainable Forestry Fund (ASFF), for $160 million—whose core funding was provided by Colonial Development Corporation, a British development company. Since 2019, GEF's timberland assets in Africa have been managed by Criterion Africa Partners, a spin-off of GEF.[21]

Some of these initial international timberland investments made by US-based TIMOs were spectacularly successful, as measured by investment returns from early acquisitions in New Zealand, Australia, Chile, Brazil, and Uruguay. But since then, exchange rate volatility, shipping costs, and intense international competition in the wood markets have led to more mixed results in some regions. More importantly, investments in emerging markets have unforeseeable market and governance risks.

For example, the Harvard Management Company (HMC) was blindsided by a middleman in Romania who was later convicted for conspiring with sellers to artificially inflate the price of timberland that HMC bought in exchange for a series of bribes, including $1.3 million, a 2007 trip to the Canary Islands, and a Chrysler Sebring car. This was part of the reason that HMC sold its timberland and other assets in Romania.[22] In 2017, HMC announced that it would write down some $1.1 billion in natural resource assets, some of which were timberlands it had invested in overseas, thereby dragging down returns.[23] This action indicates a possible overvaluation problem in the natural resource assets owned by HMC. There were also protests on the Harvard University campus regarding HMC's management of natural resource assets overseas.

Investments Made by Non-US-Based TIMOs outside the United States

The Rise of TIMOs outside the United States

As the concept of institutional timberland investment gained acceptance, there was a dramatic increase in the number of TIMOs in the United States and elsewhere to facilitate and manage these investments. After all, the TIMO business model is not complicated, at least in countries where sufficient silviculture, management, finance, and marketing capacity exist. To set up a TIMO in these countries, one needs to have a few registered financial/investment advisers and foresters who together can raise funds and handle fiduciary, investment, and property management responsibility, who can communicate in the specialized language used by institutional investors and forestry professionals with each other and with their clients, and who can meet the demands of institutional investors with the supply of forestry assets. Of course, the hard part is generating attractive investment returns based on round-trip performance—some TIMOs never got off the ground, and others simply failed after a few years.

Not surprisingly, a few people who worked in the pioneering TIMOs (Hancock, RII, First National Bank of Atlanta/FIA/Wachovia) went on to set up their own firms in the United States and other countries. It just so happened that the US forest products industry was in its last stage of timberland sales on the supply side, which fueled the growth of the TIMO industry in the 1990s and early 2000s. In emerging markets where there are no strong forest products markets or established forestry consulting partners, where silvicultural and forest management capacities are lacking, and where land titles are less certain, setting up a TIMO there or having an existing TIMO elsewhere to perform investments there would be more challenging. I will note some of these challenges in specific cases in various countries below.

The TIMO business model has spread to multiple developed and developing countries since the 1990s. Again, some of the executives who set up these TIMOs worked for US-based TIMOs as employees or consultants. Other entrepreneurs have created their own firms independently in response to the divestment of industrial timberlands in their regions, even though they might have learned from the experience of TIMOs and industrial timberland divestments in the United States. Still, small "boutique" TIMOs have existed in many countries since the 1970s or earlier, representing the third method of TIMO creation: an organic, completely new invention in response to country-specific needs that had nothing to do with the rise of TIMOs in the United States in the last three-plus decades.

An example of the first type of TIMO creation is Australia-based New Forests Pty Limited, whose founder, David Brand, worked at Hancock Natural Resources Group, which owns Hancock Timber Resources Group. Representatives of the second form of TIMO creation include Tornator in Finland and Bergvik Skog in Sweden, both of which were large TIMOs created in a way similar to US-based TIMOs—the divestment of timberlands by forest products firms in these countries.

Examples of the third form of TIMO creation include the aforementioned Forestal Caja Bancaria and Caja Notarial in Uruguay, Fountain Forestry in the United Kingdom, Société Forestière de la Caisse des Dépôts in France, International Woodland Company in Denmark, several real estate companies that have timberland funds (such as AXA Real Estate Investment Managers in France and Catella Real Estate AG in Germany), and Teak Resource Company (ex Floresteca). Teak Resource Company has directly owned timberlands since 1994 and has received investments from US and European institutional timberland funds.

The remainder of this section presents a few examples of these three forms of TIMO creation outside the United States, highlighting that timberlands have increasingly been realized as a viable alternative investment asset class on all continents that grow trees; that new opportunities arose for TIMOs from industrial timberland divestments in northern Europe, aggregation of private timberlands in western Europe, and restitution of private timberlands in eastern and central Europe; and that existing TIMOs and their investors went on to new frontiers in eastern Europe, Africa, Asia, and Latin America. These examples are illustrative and represent only a small fraction of the TIMOs in these regions today.

* * *

Founded in Sydney, Australia, in 2005, New Forests had about US$5.0 billion in assets under management and nearly one million hectares globally as of December 31, 2020. It focuses on managing assets with both production and conservation values.[24]

New Forests has three main areas of investment: Australia and New Zealand, Southeast Asia, and the United States. It has three funds for Australia and New Zealand, one Tropical Asia Forest Fund, and an investment vehicle that supplies the California carbon market with high-quality forest carbon offsets. The Tropical Asia Forest Fund, closed with US$170 million in 2013, is one of the large dedicated institutional forestry funds in Southeast Asia, focusing on forests in Malaysia, Indonesia, and Vietnam.

* * *

Tornator is a leading European company specializing in sustainable forest management. As of December 31, 2020, it had about 715,000 hectares, mostly in Finland, and the remainder in Estonia and Romania. The current Tornator was created in 2002 when Stora Enso divested its forestlands in Finland. However, the name Tornator has a long history in the Finnish forest industry. Enso-Gutzeit acquired a forestry company called Tornator in the 1930s, and the name was resurrected in 2002.

Similarly, Bergvik Skog was created in 2004 when Stora Enso and another Swedish forest products company, Korsnäs, exited the direct ownership of timberlands in Sweden. Bergvik Skog acquired 1.54 million hectares from Stora Enso and 320,000 hectares from Korsnäs in the mid-2000s.

Unlike forest products companies in the United States, Stora Enso, when it decided to divest its timberlands, continued to hold 41 percent of Tornator shares and 48.99 percent of Bergvik Skog shares. Korsnäs owns 5 percent of Bergvik Skog shares as well. The remaining shares of Tornator and Bergvik Skog belong to pension insurance companies, banks, and other institutional investors. The reason? Forestlands in Europe are valued based on the fair market value principle. Even though some forest products companies might want to deploy capital to other assets and sell forestlands from time to time, their forestlands are not undervalued, unlike under GAAP in the United States.

Interestingly, in November 2017, Stora Enso announced its intent to acquire forest assets through the restructuring of Bergvik Skog by transforming its 48.99 percent of Bergvik Skog shares to 69.8 percent of the shares in Bergvik Väst, a subsidiary of Bergvik Skog that holds 83 percent of Bergvik Skog's forestland assets in Sweden.[25] When this was done in May 2019, Stora Enso's forest holdings in Sweden increased to 1.4 million hectares, of which 1.15 million hectares were productive forestland (more than Stora Enso's share of Bergvik Skog's productive forestland of 936,000 hectares before). Stora Enso has since put these forestlands into a wholly owned subsidiary and thus reintegrated forestlands with its manufacturing business. Subsequently, Bergvik Skog and Tornator could no longer be considered TIMOs and are not listed in table 5.1.

* * *

Dasos Capital Oy is a TIMO based in Helsinki, Finland. Founded in 2005, it had a combined total volume of funds and assets of US$1.1 billion at the end of 2020. Most Dasos investments are in Europe, especially Finland, Portugal, Estonia, and Ireland. Dasos also had a small investment in an acacia plantation in Sabah, Malaysia, in 2010.

The creation of Dasos was initially supported by a few institutional investors as well as European Investment Bank through equity investments. Two other European TIMO firms—Arbaro Fund (Luxembourg based, managed by FIM Asset Management S.à r.l. and advised by Arbaro Advisors GmbH) and Althelia (based in the United Kingdom)—also received equity investments from European Investment Bank. These investments from European Investment Bank in the form of equity to TIMOs were quite rare, as they had previously been mostly in the form of loans. The bank's motivations were to promote private investment in scaling up the management of parcelized forestlands and build up natural capital in Europe and elsewhere. More specifically, it wanted to

- complement and extend lending activity to the forestry sector,
- help consolidate scattered forest ownership,
- promote more efficient silviculture and harvesting operations,
- seek attractive returns with value creation driven by biological growth, and
- capture upside potential from carbon and other nontimber markets.[26]

These were very similar to the motivations of other institutional timberland investors. It could be said that these three TIMOs rose with the help of European Investment Bank. Thus, they represent a different motivation for forming TIMOs in Europe: they were created partly in response to the demand and opportunities to consolidate small-scale and passive private forest ownership units in western Europe and make them viable management units.

Because all countries in western Europe have their own growing conditions and regulations, many TIMOs there specialize in one or two countries. The assets under management of these "boutique" TIMOs in Europe are often less than US$200 million and are consolidated from various small landowners. There are significant differences between the United States and Europe with respect to the sources of TIMO-managed forest assets—industrial timberland divestment in the United States versus consolidation of small private forestlands in Europe—and the size of TIMOs. There has also been institutional timberland investment in Ireland, Poland, Spain, and the United Kingdom. International Woodland Company estimated that as of 2016, about US$7 billion in timberland investments were under institutional ownership in Europe.[27]

* * *

The International Woodland Company (IWC) was established by two Danish institutional investors as a timberland investment adviser in 1991, when the Danish Teachers' Fire Insurance Company (Lærerstandens Brandforsikring, LB) and the Danish Pension Fund for Engineers (Danske Ingeniørers Pensionskasse, DIP) saw forest investments as a promising asset class. Especially, LB and DIP saw an opportunity to buy newly planted Sitka spruce forests in Ireland, creating a joint venture called DanWood. To manage their initial investment of DKK 45 million, spread over 2,200 hectares of land in two different locations in Ireland, they hired Otto Reventlow as IWC's CEO to oversee the investment in close collaboration with Woodland, an Irish forestry management company.

IWC has served as the adviser of other institutions in Denmark for investments in the form of primary funds, secondaries, and coinvestments. Over the years, IWC has advised and managed forestry investments on behalf of twenty clients with cumulative commitments totaling US$4.1 billion, spread over fifty funds and twenty different TIMOs in Europe, North America, Oceania, and Latin America.[28] Although its main service is long-term expert consulting, IWC had a Timberland Partners I K/S with US$230 million from institutional investors in 2014.

Thus, IWC has gradually played three roles—as a TIMO, a manager of a "fund of funds," and mostly as a consulting firm. One critical factor that facilitated the growth of IWC is that its clients, mostly Danish public pensions, pay Danish taxes, which can offset FIRPTA. Therefore, they do not suffer relative returns when investing in the United States.

* * *

One of the largest non-US-based TIMOs is Stafford Capital Partners. As of 2019, Stafford managed timberland investment in the United States, Canada, Brazil, Uruguay, Chile, Central America, New Zealand, and Australia. Interestingly, Stafford has developed a track record and is a leader in secondary markets. In a news release in November 2016, Stafford stated that over the past twelve months it had completed US$497 million in timberland investment across ten transactions including six secondary transactions, two separate account mandates, and two primary timberland fund commitments.[29] Stafford had also been the manager of Phaunos Timber Fund, which was traded on the London Stock Exchange before it was taken private in 2018.

Many non-US TIMOs started out as regional specialists. IWC and Dasos started in Europe, New Forests in Australia, Teak Resource Company in Brazil, and GEF/Criterion Africa Partners in Africa. Investors liked the idea of a local manager with apparently better local knowledge despite GFP and Hancock having strong regional teams.

Pushing the Frontiers of Institutional Timberland Investment

Figure 7.4 shows a map of the mature, intermediate, emerging/frontier markets for timberland investment from New Forests' perspective. The mature markets include North America, western Europe, and Oceania. The intermediate markets include Brazil, Chile, central and eastern Europe, Malaysia, Mexico, and South Africa. The emerging markets cover China, India, Russia,

Figure 7.4. Global timberland markets by maturity

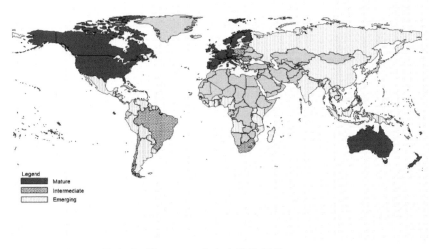

Source: New Forests, Timberland Investment Outlook 2013–2017

and most countries in Southeast Asia as well as countries in East and West Africa. This perspective is shared among many TIMOs and their institutional investors.

As noted earlier, some TIMOs in the United States and Europe went beyond North America and Europe to Oceania and Latin America, seeking high risk-adjusted returns from tree growth there even before industrial timberland divestments were completed in the United States. Then, more TIMOs expanded to eastern Europe, Africa, and Asia. Table 7.1 provides a list of some active boutique TIMOs on these continents that are not listed in table 5.1.

Table 7.1. Active boutique TIMOs by continent and country

	Headquarters	Countries invested in	
Africa			
Criterion Africa Partners	US	South Africa, Tanzania, Swaziland, Uganda, Gabon	https://www.criterionafrica.com/
Form International	The Netherlands	Ghana, Tanzania	http://www.forminternational.nl/
Green Resources A/S	Norway	Mozambique, Tanzania, Uganda	http://www.greenresources.no/
Miro Forestry	UK	Ghana, Sierra Leone	https://www.miroforestry.com/
The New Forests Company	South Africa	Rwanda, Uganda, Tanzania	http://newforests.net/
Asia			
SilviCapital	Sweden	Laos	http://www.silvicapital.com/

	Headquarters	Countries invested in	
Europe			
Aquila Capital	Germany	Romania, Finland, UK	http://www.aquila-capital.de/en
AXA Real Estate Investment Managers	France	Finland, France	https://www.axa-im.com/en/
Catella Real Estate AG	Germany	Finland, Argentina, Panama, US	https://www.catella.com/
Euroforest AB	Sweden	Lithuania, Latvia, Estonia	http://en.euroforest.se/
Foran Real Estate	Latvia	Latvia	http://foranrealestate.lv/
Forest Value Investment Management	Luxembourg	Romania	http://www.fvim.lu/
Global Forestry Capital SARL	Luxembourg	Brazil	http://globalforestry.com/
International Woodland Company	Denmark	US, Ireland, Cambodia	http://www.iwc.dk/
Irish Forest Unit Trust	Ireland	Ireland, US	http://www.iforut.ie/
Irish Forestry Funds	Ireland	Ireland	http://www.irish-forestry.ie/
Latifundium Management GmbH	Germany	US, Panama, Argentina, Uruguay, New Zealand	www.latifundium.de
QUADRIS Environmental Forestry Fund	UK	Brazil	http://www.quadris-funds.com/
Latin America			
AF (Bosques del Uruguay)	Uruguay	Uruguay, Brazil, Chile	http://www.bosquesdeluruguay.com/
Amata	Brazil	Brazil	http://www.amatabrasil.com.br/
Arbaro Fund	Germany	Paraguay, Uruguay	http://www.arbaro.lu/ Steffen Klawitter, s.klawitter@ finance-in-motion.com
Atico Florestal FIP	Brazil	Brazil	http://www.atico.com.br/
Central America Timber Fund	Luxembourg (retail investors)	Costa Rica	https://www.catf.lu/
Compañía Agrícola de la Sierra	Chile	Colombia	Juan Francisco Munoz (juanfcomunoz@mcondor.cl)
Copa Investimentos	Brazil	Brazil	http://www.copainvest.com.br/
Forest First Colombia S.A.S.	Colombia	Colombia	http://www.forestfirst.com/
Futuro Forestal	Panama	Nicaragua, Panama	http://futuroforestal.com/
Lacan Florestal FIP	Brazil	Brazil	http://www.lacaninvestimentos.com.br
Orion Capital	Chile	Chile, Colombia	http://www.orioncapital.cl/
Proteak Uno	Mexico	Mexico, Colombia, Costa Rica	http://www.proteak.com/index.php/en/; traded on the Mexican Stock Exchange
The Forest Company Ltd.	Guernsey, UK	Brazil, Colombia	http://theforestcompany.se/
Uruguayan Pension Fund	Uruguay	Uruguay	
VBI Timberland	Brazil	Brazil	http://www.visionbrazil.com/
Oceania			
OneFortyOne Plantations Holdings	Australia	Australia, New Zealand	https://onefortyone.com/

The rise of regional TIMOs in South America is associated with US and European institutional investors there. As Table 5.1 shows, two Brazil-based TIMOs are on the list of the top twenty-five TIMOs in the world. On the other hand, the expansion of TIMOs to eastern and central Europe benefited from the restitution of private forest assets to their pre-Communist owners after the collapse of the Berlin Wall. Many of these "new" forest owners were often "absent" from their forests and their livelihood did not depend on forestry, so they were looking for managers to manage their forests. The number of small holdings has also increased as a result of restitution and privatization. It was only after Estonia, Hungary, Latvia, Lithuania, and Poland joined the European Union in 2004, and Romania in 2007, that many TIMOs went to these countries, including the US-based Harvard Endowment and GreenWood Resources, and many Europe-based TIMOs such as Bergvik Skog (in Latvia), and Tornator (in Estonia and Romania).

In theory, the restitution process was straightforward: claimants submitted their evidence of private ownership, and the authority (either a restitution commission or court) approved, adjudicated, and settled disputes, if any. In practice, this process was more complicated and subject to fake evidence and corruption. For example, in Lithuania, a restitution right could be transferred to another person, and preference of restituted forestland given to similar land in the same place. If similar land could not be found in the same place, a piece of land elsewhere with the same value was offered. This gave people who did measurement and valuation an opportunity to buy and accumulate forestlands from those who had restitution rights, because the former had better information. In Romania, it is observed:

> For the stakes of the final stage of restitution (Law 247/2005) were much higher ... the claimants had to produce solid evidence for their requests. Perfect time for a special type of stakeholder to show up on the scene: Middlemen—rich businessmen, with a good grasp on the juridical system, bold, tenacious and able to produce, if needed, fake documents, resembling the ownership titles issued seven decades ago, yellowed by time ("a microwave oven and some experience in cooking the paper were enough," confessed the interviewed private owner).[30]

So much fraud occurred that a Romanian government audit showed that one-fifth of the cases of restitution to private owners of forestland confiscated by the Communist regime in Romania were illegal.[31] Not surprisingly, some TIMOs and industrial timberland owners who bought forestlands in Romania had to write off a portion of their assets there and suffered a loss. On the other hand, if due diligence had been performed, one could have purchased promising assets in Romania and other eastern and central European countries at a hefty discount. This lack of information and transparency is a market imperfection that has been exploited by some bold TIMOs.

* * *

As noted earlier, the largest TIMO that invested in Africa was the US-based Global Environment Fund (GEF) or the following Criterion Africa Partners. Yet the first TIMO that invested a large amount of institutional funds in African countries other than South Africa was Green Resources A/S, a private Norwegian company financed by Norwegian investors, in Mozambique in 1995.

In 2005, Global Solidarity Forest Fund AB, an investment entity started by the Church of Sweden's Diocese of Västerås, invested in Mozambique first and then expanded to Tanzania and Uganda for a total of some 41,000 hectares of eucalyptus plantation. Other investors in this fund included the Harvard University Endowment and a Danish pension plan (ABP) advised by the International Woodland Group. In 2014, the assets of Global Solidarity Forest Fund AB in East Africa were acquired by Green Resources A/S. As of early 2020, Green Resources had 38,000 hectares of forest plantation assets in East Africa and Mozambique valued at US$147 million.[32]

The challenges for TIMOs investing in Africa (with the exception of South Africa) are threefold: low returns because of lack of infrastructure, high establishment and transportation costs, and insufficient scale; structural barriers such as land tenure issues, stakeholder relations, inadequate silvicultural knowledge, and downstream processing capacity; and technical and capacity issues such as lack of management expertise and unproven planting materials and silvicultural practices.[33] One European-based TIMO had to write off some assets in an African country when it found out that the eucalyptus trees it had planted did not grow well after two years. Apparently, there was a hardpan 1.5 meters underground that the eucalyptus roots could not penetrate after two years. This TIMO did hire a soil expert to test the soil, but only to 1.2 meters belowground. In Mozambique, Green Resources is preparing to turn hundreds of thousands of acres of concession area in Lurio province back to the government primarily because it lacks the funding to develop plantations and has concerns about reputational risk.[34] Additionally, to solve the problem of the lack of local processors, many TIMOs have tried to integrate vertically into manufacturing as they develop plantations in Africa.

As a result, the African Development Bank estimated that there was less than US$1 billion in investment in Africa's forest sector in the whole decade prior to 2017. Private high-net-worth investors invested about US$400 million; the Development Finance Institution provided US$280 million, primarily to fund the construction and operation of processing facilities; and institutional investors (through TIMOs) invested about US$200 million (outside South Africa).[35] Compared to the $70 billion in assets under management by TIMOs worldwide, the amount of institutional timberland investment in Africa was rather small.

* * *

Asia has the largest population in the world and thus a large and fast-growing market for forest products, and the forest industry in Asia is quite advanced. Unlike in North America and northern Europe, industrial timberland divestment did not happen as widely in Asia. Nonetheless, reform in forest tenure (privatization of forestlands to some degree) in China and several countries in Southeast Asia in the last forty years has offered an opportunity for some TIMOs to enter this region. The main challenges for TIMOs here are high transaction costs of business operation because of forest tenure, parcelization of forestlands, regulations, and the high maintenance costs of forest assets. Compared to other regions, Asia has perhaps the least amount of timberland managed by TIMOs, especially on a per capita basis. National and family forest ownership still dominates here.

* * *

Figure 7.5 shows the view of one TIMO—Dasos—regarding the destinations of institutional timberland investment as measured by market maturity, from mature to intermediate and emerging/frontier markets in relation to discount rates. Much like UBS RII two decades ago (fig. 7.3) and many other TIMOs nowadays, Dasos sees the discount rate or the required rate of return for timberland investment rising from mature (such as Sweden) to intermediate (such as Uruguay) to frontier (such as Mozambique) markets. If there is a universal timberland asset that covers all global timberlands and if all risk is accounted for—such as forest growth, transaction costs, market conditions, languages, exchange rates, and political stability in all countries—all

Figure 7.5. Timberland investment discount rate as a function of market efficiency

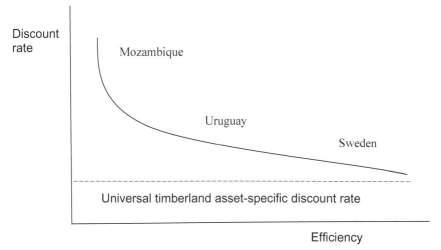

Source: Dasos (2019).

timberland investments should have a return close to a "universal timberland asset specific discount rate." Conceptually, this universal timberland asset-specific discount rate is similar to the risk-free rate plus a market rate for timberland investment in the United States. Anything above it would be compensation for additional idiosyncratic project risk.

Publicly Traded Timberland and Forest Products Funds

There are a few publicly traded international timberland and forestry funds on various stock exchanges. Two of them are or were purely international timberland funds, and others are global timberland and forest products funds. The performance of the publicly traded international timberland funds has been mediocre at best. Thus, as of 2020, there was no viable publicly traded option for pure international timberland investment.

The first publicly traded timberland fund was Phaunos Timber Fund Limited (Phaunos), which was registered in Guernsey, United Kingdom, on December 20, 2006, and whose shares were initially traded on the Alternative Investment Market and then the London Stock Exchange. With an initial investment of US$115 million in 2006, the fund aimed to provide shareholders with attractive long-term returns through exposure to a global portfolio of timberland and timber-related investments. As of December 31, 2016, Phaunos had approximately US$299.4 million in assets in New Zealand,

Table 7.2. Assets of Phaunos Timber Fund Limited at the end of 2016

Country	Name	Share of ownership	Size	Valuation
New Zealand	Mataraki, a joint venture with Rayonier	35%	169,000 hectares	US$130.7 million (2014)
US	National Timber Partners	Minority	---	US$2.1 million
	Greenwood Tree Farm Fund	Minority	14,000 hectares	US$36.2 million
Uruguay	Aurora Forestal	Minority	19,000 hectares with a sawmill	US$31.2 million
	Pradera Roja	100%	9,200 hectares	US$23.3 million
Brazil	Eucateca	100%	16,000 hectares	US$34.3 million
	Mata Mineira	100%	19,000 hectares	US$49.7 million

Source: Phaunos Timber annual report, 2016.

North America, and South America, all through joint ventures with local or international partners.[36] The fund was initially managed by FourWinds Capital Management before 2014 and by Stafford Capital Partners afterward. Table 7.2 presents the assets of Phaunos at the end of 2016.

After the loss of a continuation vote at the 2017 annual general meeting on June 19, 2017, Phaunos's investment policy was amended to permit an orderly realization of its assets. In 2018, the fund started to execute its shareholder-mandated asset sales process and received a final offer from Stafford of US$0.52 per share, or some US$259 million for the entire company, which it eventually accepted. Phaunos was delisted on December 6, 2018.

The other publicly traded timberland fund was Cambium Global Timberland Limited, which was registered in Jersey, United Kingdom, on January 19, 2007. The company initially raised £104 million and invested in Australia and in Texas and Hawaii in the United States. It initially appointed Cogent Capital Asset Management LP as its asset manager, and its shares were listed on the Alternative Investment Market of the London Stock Exchange on March 6, 2007. However, after a few years of unsatisfactory performance, it sold many of its assets under pressure from its shareholders.[37] As of April 30, 2016, its total assets under management were only £19.3 million, which included two plantations in Brazil and one in Hawaii. In 2020, it was selling or renting all its assets.

There were also a couple of failed publicly traded pure timberland companies outside the United States. The first was Evergreen Forests Limited. Listed on the New Zealand Stock Exchange on September 10, 1993, it owned blocks of forests on the west coast of South Island and in central and northern North Island in New Zealand. However, after years of mismanagement and poor returns, its shareholders voted to liquidate its assets. It was delisted on July 21, 2006. The other such company was Sino-Forest Corporation, which was traded on the Toronto Stock Exchange between 1995 and 2012. Sino-Forest Corporation claimed to have large timber assets in China, but this claim was later found to be questionable and unsubstantiated. After the collapse of Sino-Forest Corporation's stock price in 2011, the Toronto Stock Exchange delisted its shares on May 9, 2012.

There are two US-based public timberland and forestry funds. The first is the iShares Global Timber & Forestry exchange-traded fund (ETF), which has been traded on the NASDAQ (trade symbol: WOOD) since June 24, 2008. The objective of this fund is to provide shareholders with exposure to global equities in or related to the timber and forest industry. The ETF had a net asset value of $301 million as of December 31, 2020. The top ten companies in this ETF are presented in table 7.3. Of the top ten shareholdings, pure-play timberland companies represented only 25 percent of the equity in the fund.

The other US-based timberland and forestry fund is the Invesco MSCI Global Timber ETF (trade symbol: CUT), which has been listed on the New York Stock Exchange since November 9, 2007. This was the first US-listed global timber ETF to offer retail investors access to the global timber and

Table 7.3. Top ten holdings in the iShares Global Timber & Forestry ETF as of December 31, 2020

Name	Sector	Country	Weight (%)
Svenska Cellulosa B	Materials	Sweden	8.29
Weyerhaeuser Company	REIT	US	8.04
PotlatchDeltic Corp.	REIT	US	7.34
Rayonier Inc.	REIT	US	7.29
West Fraser Timber Ltd.	Materials	Canada	6.91
Holmen Class B	Materials	Sweden	4.18
Smurfit Kappa Group	Materials	Ireland	4.01
Westrock	Materials	Japan	3.97
Sumitomo Forestry Ltd.	Consumer discretionary	Finland	3.92
UPM-Kymmene	Materials	US	3.92

Source: iShares Global Timber & Forestry ETF, accessed January 3, 2021, https://www.ishares.com/us/products/239752/ishares-global-timber-forestry-etf.

Table 7.4. Top ten holdings in the Invesco MSCI Global Timber ETF as of December 31, 2020

Name	Weight (%)
Weyerhaeuser Co.	5.30
UPM-Kymmene Oyj	5.30
Avery Dennison Corp.	4.79
Packaging Corp. of America	4.76
International Paper Co.	4.57
Mondi PLC	4.54
Smurfit Kappa Group PLC	4.54
Amcor PLC	4.53
Westrock Co.	4.50
Stora Enso Oyj	4.43

Source: Invesco MSCI Global Timber ETF, accessed January 3, 2021, https://www.invesco.com/us/financial-products/etfs/product-detail?audienceType=Institutional&ticker=CUT.

forest products market. As of December 31, 2020, it had a net asset value of only $90.8 million. The top ten holdings in the fund are presented in table 7.4.

Finally, there is Pictet-Timber Fund (ISIN: LU0340559557), which has been trading on the Luxembourg Stock Exchange since September 9, 2008. Although it is called a timber fund, most of its top ten holdings in 2020 were forest products companies around the world (table 7.5). As of July 31, 2020, it had an asset value of €384 million.

Table 7.5. Top ten holdings in Pictet-Timber Fund as of July 31, 2020

Name	Weight (%)
Weyerhaeuser Co.	7.69
PotlatchDeltic Corp.	5.88
West Fraser Timber Co. Ltd.	5.58
Rayonier Inc.	5.24
Interfor Corp.	4.17
Canfor Corp.	4.08
Stora Enso Oyj-R shares	2.78
Svenska Cellulosa AB Sca-B	2.35
Norbord Inc.	2.22
Louisiana-Pacific Corp.	2.15

Source: Pictet Asset Management, accessed December 20, 2020, https://www.investors-trust.com/pdf/pic/eng/PICTE.pdf.

These three ETFs give investors indirect exposure to timberland investment through a basket of global equities that own or lease forestland, harvest the timber for commercial uses, and produce wood-based products. Again, unlike timberland REITs in the United States, they are not pure timberland players. For investors who want to access pure-play timber assets, the only option is to purchase the top four or five companies (WY, RYN, PCH, CTT, ADN) on a capitalization-weighted basis.

Implications for Global Forest Sustainability

The implications of institutional timberland investment through TIMOs for global forest sustainability are fivefold. On the positive side, TIMOs prevent or slow down the parcelization of industrial timberlands when these timberlands are put on markets, or amalgamate small-scale forest ownerships to make them reasonably large management units. The scale of TIMOs provides economies that can reduce the unit cost and improve the efficiency of forest management globally. The aggregation of timberland management by some TIMOs in Europe and Asia might have made industrial forestry a viable business there.

Second, TIMOs improve the asset allocation efficiency of financial capital and timberlands. At a minimum, if TIMOs can buy timberlands from their existing owners, it is an improvement in terms of Pareto efficiency, which is defined as an economic state where resources cannot be reallocated to make one individual better off without making at least one individual worse off. TIMOs can offer a higher price for an "undervalued" asset because they have better information on inventory or more reliable growth-and-yield information, because they can create additional value through management, and/or because they can raise capital at low cost.

Third, TIMOs can often further improve asset management efficiency by realizing the full potential of timberland for land appreciation, timber growth, and price appreciation because of deep knowledge of local markets and technical expertise. Furthermore, TIMOs are more likely to introduce added value to the timberland from payments for ecosystem services including carbon markets, and HBU. As for converting forestlands to other uses, there is no evidence that TIMOs have higher or lower rates of land use conversion than industrial and other private forest owners.

Finally, and from the demand side, the best TIMOs respect sustainability principles and comply with independent third-party certification schemes. Adherence to broadly defined environmental, social, and governance (ESG) metrics is a must to comply with institutional investor ethics. Thus, they maintain forest certification and sound forestry practices in managing their timberlands. Their positive actions can catalyze other forest landowners to follow suit.

The only drawback of increasing international timberland ownership seems to be the possible mismatch of interests between absentee owners and the local labor force and communities, although this can be partially mitigated through local property managers. As Conner Bailey and his colleagues have shown, concentrated and absentee ownership of land can create conditions of an internal colony, where key decisions are made by powerful outside forces supported by local elites. In addition, both concentrated and absentee ownership of timberland are negatively associated with quality of life in rural Alabama as measured by educational attainment, poverty, unemployment, food insecurity, eligibility for free or reduced-price lunch at public schools, Supplemental Nutrition Assistance Program participation, and population density.[38]

So, the impacts of global institutional timberland investment are mostly positive, although some sociologists worry that concentrated and absentee landownership may have a negative impact on local communities. We can hope that the investment decisions being made today in pension fund and endowment boardrooms will not bring added tensions between timberland as a financial asset and its physical and social properties.

CHAPTER 8
Promises and Prospects

Timberland ownership has changed substantially in the United States and elsewhere. Large tracts of industrial timberland in the United States accumulated by vertically integrated forest products companies for nearly the last century were either sold, mostly to institutional investors, or converted to publicly traded timberland REITs. If the accumulation of industrial timberlands in the United States was a response to a natural barrier that induced imperfect competition, the shift to institutional timberland ownership is mostly the result of market imperfection caused by institutional factors—changes in corporate management philosophy, differing valuation metrics, relative tax efficiencies, and accounting rules—as well as market dynamics. Institutional timberland investment in other countries, although not as large in scale as in the United States, rose with more nuanced motivations, including privatization of public forest estate, industrial timberland divestment, augmentation of small forest tracts, and restitution of private lands to their previous owners.

Now the low-hanging fruit of converting industrial timberlands to institutional ownership has largely been picked in the United States and elsewhere. And so far, institutions and others have developed very few new forestry assets, plantation or otherwise. So, most transactions are trades of extant forests among mostly extant investors. It is time to look ahead to future trends in institutional timberland investment around the world. Before doing that, I provide a summary of the rise of institutional timberland investment and address the question of whether the promises of such investment have been kept in the last few decades.

The Rationale and Promises Mostly Kept
The Rationale

The historical accumulation and recent sales of industrial timberland in the United States and elsewhere as well as the rise of institutional timberland ownership were driven first by industrial entrepreneurs and then by institutions pursuing high returns from imperfect timber and timberland markets.

The bulky nature of raw timber means that the transportation costs of timber for forest products manufacturing are high. Fueled by increasing demand for their products in much of the twentieth century, manufacturing plants have grown larger in terms of production capacity and capital investment in one location and subsequently needed larger volumes of timber from surrounding areas. This gives forest products manufacturers an opportunity to explore locational monopsonistic or oligopsonistic rents by owning some timberlands not far from their mills. Industrial timberlands reduce competition and the prices of timber on open markets, enhancing the profitability of manufacturing plants. Owning timberlands also provides additional benefits to corporate managers because income generated from timberlands can be used to stabilize corporate earnings and thus give them job security.

In the United States, industrial timberlands prior to 1900 might have come from the government. In the twentieth century, however, they came mostly from farmers, who were often not informed about the value of their timberlands, did not know how to project the volume and value growth of their forests, and often did not measure timber inventory. Furthermore, some of these farmers would sell not only their forestlands but also their agricultural lands if the offering prices were higher than or comparable to the prevailing prices of agricultural lands. In economic terms, this is to say that they sold their agricultural lands when the offering prices were higher than their opportunity costs. When agricultural products experienced a period of cyclically low prices in the United States, such as in the late 1950s and early 1980s, some farmers sold their farmlands at depressed prices and moved to cities and suburbs, and these farmlands have since been converted to forestry uses.

Because forests have a long age to maturity, forestry is a capital-intensive business. I have calculated that the capital-to-output ratio for forestry likely reaches fifteen to one for softwood forests and even twenty to one for hardwood forests in the US South. The capital-to-output ratios for forestry in other parts of the country are higher.[1] Thus, no major forest products firms have reached close to 100 percent self-sufficiency in timber supply. At its peak in 1987, the national average timber self-sufficiency rate was about 30 percent.

Gradually, the development of transportation networks, tree-growing technology, and silvicultural techniques, combined with the high cost of capital faced by forest products companies, raised the question of whether owning capital-intensive timberlands might be too costly, especially if timberlands were treated merely as a source of raw timber supply. It did not help that under GAAP, industrial timberlands were undervalued, resulting in some industrial firms facing hostile takeover bids from financiers and corporate raiders who wanted to buy them and then sell them in pieces. Even without hostile takeover bids, the gap between the market value and the book value under GAAP grew large for some timberlands, and forest products firms were better off even if they sold them at a discount to their market value to improve returns on their capital and balance sheet. Furthermore, the tax structure of most forest products firms was not efficient, and the capital gains tax treatment of timber income from industrial timberlands disappeared in 1986. Naturally, some insiders and shareholders of these companies as well as some academicians and financial analysts on Wall Street started to recommend that these firms sell some of their timberlands.

Since industrial timberlands are normally in large tracts and are expensive, the only logical buyers are institutional investors, especially tax-exempt institutions that have a long-term perspective and a lower cost of capital than tax-paying institutions. Some of these institutional investors had a dismal record with their previous investments and were asked to diversify their investment portfolios under ERISA of 1974. At this juncture, some TIMO pioneers started to connect the two—industrial timberland owners as sellers and institutional investors as buyers—and became the advisers of the newly emerged institutional timberland owners and managers of their timberland assets.

The TIMO pioneers and early practitioners understood that timberland should be attractive to institutional investors, as anecdotal evidence at the time showed that timberland investment historically generated attractive returns, displayed low risk in terms of volatility, and had high potential for asset diversification and inflation hedging. They aggressively promoted timberland to institutional investors and others, and some of them succeeded. Also attractive was the historical opportunity to buy low. In the end, both the sellers and buyers of industrial timberlands struck deals that benefited both sides. For the sellers, the sales increased the shareholder value of major US forest products companies. As for the buyers, they obtained a stable asset class and have since captured excellent risk-adjusted returns.

Market conditions such as recessions, high debt, and high cost-of-capital environments expedited the sale of industrial timberlands. And those that did not sell chose to treat their timberlands as profit centers or set up timberland MLPs over which they still had some control. As the regulatory environment improved, REITs became more attractive than MLPs. Thus, most timberland MLPs and some timberland-heavy forest products companies converted themselves to timberland REITs. Such conversion was completed in 2010, a year that also marked the nearly complete transfer of industrial timberlands to institutional ownership in the United States.

The Structure, Forest Management, and Global Reach of TIMOs and REITs

Some industrial timberlands were sold with a wood supply agreement that provided some assurance of wood availability to sellers and a degree of protection to institutional investors. Nonetheless, institutional timberland owners did not tie their timber to a mill and could make timber harvesting decisions based on market conditions. Thus, in theory, the timber supply elasticity (that is, the price responsiveness of timber) of institutional owners—both TIMOs and timberland REITs—differs from that of traditional industrial owners or nonindustrial private forest owners.

Normally, a timberland REIT must derive at least 75 percent of its gross income from real estate sources including timber sales.[2] As only 25 percent of its income can be from nontimber sources, it needs to harvest timber to make regular dividend payments, even in the face of weak timber markets—unless and until it is willing to reduce or suspend dividend payouts. TIMOs do not have the same constraint.

As institutional timberland ownership has a terminal time horizon, it was feared that institutional investors would not perform long-term forest management activities. However, recent empirical studies show that these institutional timberland owners do reforest and implement other sustainable forestry management practices, and because marketing conservation brings additional income, their records in forest conservation are as good as those of other landowners.

As of December 31, 2020, US-based TIMOs or TIMOs with a significant presence in the United States collectively managed about $49 billion in timberland assets, roughly 64 percent of which were in the United States, and the remainder mostly in Oceania and South and Central America.[3] In comparison, the total enterprise value of the five biggest timberland REITs traded in US markets, including one Canadian firm (Acadian Timber), was on the order of $39 billion at the end of 2020. No viable pure-play timberland investment stocks are listed outside North America, possibly because such companies need to have a certain viable size in capitalization.

In Oceania, timberland investment—including institutional timberland investment—has been fueled by privatization of government forest assets, industrial timberland divestment, dynamic market conditions, and high tree growth rates. Industrial timberland divestment was similar to what happened in the United States, although at a smaller scale. In fact, New Zealand and Australia were the first countries US-based institutional investors went to, seeking attractive risk-adjusted returns. Institutional investors then went to South America, attracted by high rates of tree growth and industry-scale forest plantations. Later, some institutional investors went to Central America, chasing high-valued hardwood forests such as planted teak and mahogany.

Europe has a long history of institutional timberland ownership. Institutional timberland investment and TIMO businesses in Europe in the modern era have benefited from divestment of industrial timberlands, amalgamation of small forestland tracts, and timberland restitution.

Because of government and communal ownership and population pressure, institutional timberland investment has grown slowly in Asia and Africa. However, demand for forest investment in Asia is rising as Asia remains a growing source of forest products demand. Africa is seen as the most prospective frontier of timberland investment, with lots of land potentially available and strong population and economic growth. Sound policies and investment conditions at the country level are highly correlated with investment opportunities in forestland. These conditions consist of political stability, secure property rights, well-functioning legal and banking systems, strong domestic consumption of forest products, a stable tax system, and an acceptable currency risk. In the near future, investments in Asia or Africa could remain limited because of land tenure and other risks. And, if the risk-free rate for investing in a country is in double digits, investors will not be compensated for taking long-term forestry risks there.

The Results: Promises Are Mostly Kept

As chapter 6 shows, institutional timberland investment via TIMOs and REITs in the United States has secured promising returns with a relatively low level of risk to date. The performance of private equity was better than that of public equity before 2001. Since then, the returns of timberland REITs have been better and more volatile than those of TIMOs. Thus, the determination of which one is better, now or in the immediate future, depends on investors' primary objective, risk tolerance, and other needs. At a minimum, the coexistence of these two alternative timberland investment vehicles with roughly the same size in terms of capitalization affords investors a choice, and a chance to discover the opportunity cost of each investment vehicle in the United States.

Under normal conditions, timberland investment should receive adequate returns from biological growth, ingrowth (or tree appreciation), land appreciation, and price appreciation (or depreciation), as noted previously.

In the future, woody biomass for bioenergy production may present a significant opportunity to add value to existing timberland assets. While related technologies are still largely in the research and pilot phase, manufacturing biofuels from wood may become commercially viable in the future. Of course, forests produce various nontimber products and environmental services, which are mostly public goods. To the extent that landowners could capture and/or securitize part of these products and services, the returns from timberland investment would be enhanced.

Timberland investments, however, are vulnerable to risks, which can include

- overvaluation, which depresses returns;
- illiquidity in TIMO-managed assets;
- natural disasters;
- a decrease in demand for timber because of housing market slowdowns and the development of the internet;
- country risk, which includes exchange rate risk.

Thus, timberland valuation is so critical that overpricing is considered the only risk by some TIMOs. Timing is also important. Investments made at the high point of industrial timberland sales in the United States around 2003–2008 have not yielded expected returns, leading some institutional investors such as CalPERS to sell their timberlands.[4] This could be a big issue in the long run if the recent poor financial performance of US-based institutional timberland investment—as evidenced in the NCREIF Timberland Index—does not improve. In fact, timberland investment will not prosper as much unless timber prices improve and ecosystem services are better achieved, thereby enhancing its financial performance relative to other investment alternatives.

The Unanswered Questions

Rising institutional timberland investment has also raised questions about these investors' land use change behaviors, market impacts, and implications for sustainability. This book provides some evidence for improved understanding of the impacts of institutional investment. Given that institutional timberland investment has emerged in just the last few decades, many of these questions have not been fully answered.

Chief among these questions is the propensity of institutional investors to convert forestland to other uses. Because the land-holding tenure of a TIMO may be shorter than that of industrial and nonindustrial private forest owners, many policy makers and stakeholders want to know whether institutional timberland owners are more likely to sell timberlands for HBU than industrial or other private forest owners, which often leads to permanent forest loss. In the coming years, there should be enough data to lay this issue to rest in the United States and elsewhere.

It is also hypothesized that institutional investors may be more interested in developing ecosystem service markets such as carbon credits. Ecosystem services are sources of additional value that can enhance returns. Are there any differences between TIMOs and timberland REITs in providing nontimber benefits? What about harvesting behaviors and the resulting timber supply elasticity in TIMOs and timberland REITs?

Then there is the subject of conservation easements. Is it a good thing to take land out of multiple use options? When this is done, for example, for the benefit of public-employee pension funds that depend on taxing citizens to fund their pension obligations, is this really the best long-term social policy? Why do citizens at this time think that placing a perpetual restriction on private property is a good idea?

Finally, will institutional timberland ownership continue to expand at the expense of private family forests and private forest industry owners in the United States and elsewhere? Or if there is a level playing field for all timberland owners such that they pay the same amount of taxes, and if the International Financial Reporting Standards are adopted in the United States, is it possible that some forest products companies could buy back timberlands they sold previously, just as Stora Enso did in May 2019? Certainly, the extent of institutional timberland holdings is subject to changes in taxes, regulations, and other institutional factors in the future.

Trends and Prospects

The rise of institutional timberland investment has demonstrated that timberlands are a viable investment asset class for institutional investors. Today, institutional timberland investment has gone global. This development may shed some light on the future of institutional timberland ownership in the coming decades.

Arbitrage and Competition between TIMOs and REITs

As stated, institutional timberland investment through the private equity route via TIMOs and through the public equity route via publicly traded REITs differs in terms of liquidity, degree of control, valuation metrics, and investment goals. The investment goal via private equity is often total returns, and periodic cash income is desirable but not a prerequisite. On the other hand, the REIT structure was created specifically to pass pretax income on to shareholders. Publicly traded timberland REITs are thus valued at a multiple of their cash yield and typically aim at issuing stable and preferably increasing dividends over time.

The pursuit of cash dividends may induce some timberland REITs to purchase timberlands that have timber near its financial maturity—the so-called cash-yielding assets. TIMOs, on the other hand, may purposefully purchase timberlands that have young forest stands and thus take advantage of the high growth rates (including ingrowth) in these stands in the pursuit of total returns. There is an arbitrage opportunity and a potential synergy between TIMOs and REITs under normal market conditions, again because publicly traded timberland REITs are valued on funds-from-operation (FFO) and TIMOs on total returns. Young properties have low FFO but possible good total returns. Thus, there could be a cycle in which REITs sell cutover lands and premature forests to TIMOs, which sell them back when these lands and forests start to yield positive cash flows.

This kind of arbitrage is illustrated in two transactions among TIMOs (FIA, Campbell Global) and Rayonier in May 2016. A news release from Rayonier revealed that it completed two separate transactions to enhance its Pacific Northwest timberland portfolio. The transactions included the net acquisition of approximately 61,000 acres of well-stocked, highly productive timberlands in Oregon and Washington from Campbell Global, and the disposition of approximately 55,000 acres predominantly of premerchantable timber in Washington to FIA. The acquisition (average plantation age of 22.4 years) complemented the age-class profile of the company's existing Pacific Northwest portfolio (average plantation age of 19.0 years), while the disposition was mostly younger forest stands (average plantation age of 12.6 years).[5] These transactions seemed to be a win-win situation for Rayonier, Campbell Global, and FIA. In the future, we may see more of these types of transactions.

Although the performance of TIMOs and timberland REITs has varied over the last thirty-four years, my hypothesis is that in the long run, the risk-adjusted performance of both tends to equalize, net significant transaction costs of 6–10 percent for timberland transactions. If not, TIMOs could buy out timberland REITs or vice versa.

It is also possible that some timberland REITs held privately by TIMOs could be spun off and become publicly traded. Hancock Timber Resource

Group already has a private timberland REIT with a much larger net asset value than the market capitalization of $456 million in CatchMark at the end of 2020. There are other private timberland REITs with a net asset value of $500 million or more. Publicly traded timberland REITs could also be bought and taken private.

Secondary Markets

Some people have described today's asset transactions among TIMOs, timberland REITs, and direct investors in the United States as a secondary market because these transactions differ from the primary market, which is characterized by the wave of industrial timberland sales. Others may consider that the managers of "funds of timberland funds," who allocate funds raised among different TIMOs, have engaged in secondary market activities. In my mind, neither is a secondary market. In the former, the transactions are transparent and done at arm's length. In the latter, the managers of "funds of funds" are aggregators of diversified assets or merely middlemen.

What I mean by a secondary market is a situation in which one investor buys the shares of another investor in an existing timberland fund. In other words, secondaries are private sales of private timberland equity interests prior to the termination of a commingled fund. The price may be equal to or close to the net asset value of the selling investor's shares in the fund. It is often substantially less. This transaction occurs when the former wants to invest in timberlands and the latter wants to exit the timberland fund—partially or completely. Again, secondary market sales refer to purchases of limited partner interests in a private equity fund.

This secondary market transaction could occur within one individual TIMO if there were a willing buyer and a willing seller for shares in an existing fund that it managed, or if both the buyer and seller happened to use the same TIMO in different funds. It also could occur between two TIMOs with different investors. In the latter case, the transaction costs could be high because information—such as the investors in a fund, the possibility and intention of an investor who wants to get out of the fund, and the maturity, past performance, and unit price of the fund—is often confidential. Unless a good relationship and trust exist between the two TIMOs, secondary market transactions between them are hard to pull off. All investors in private equity timber funds owe a duty to the other investors to maintain strict confidentiality about the fund. Otherwise, they could potentially damage that value given the highly confidential nature of information.

Thus, secondary markets for institutional timberland investments man-aged by TIMOs are very illiquid and opaque. As a result, any investor who wishes to leave a fund early can do so, but often at a discount to its net asset value. Nonetheless, since the early 2000s, specialist investors seeking sec-ondary interests for acquisition have provided liquidity in cases where one of the limited partners in a primary fund wants to exit early and other partners choose to remain.

Timberland investments find their way to secondary markets for a variety of reasons that are unique to each investor. Changes in investment strategy, portfolio rebalancing, a desire to sell underperforming assets, and the need for liquidity are a few of the many reasons. Secondary markets for institutional timberland investors benefit both sellers and purchasers of these interests. For purchasers, acquiring a secondary interest allows them to gain access to known managers with strong performance records and to high-quality and tightly held assets.[6]

Stafford Capital Partners is the leading TIMO in secondary markets for timberland, with 114 secondary transactions in the fifteen years before 2020. As of 2019, 67 percent of the US$2.4 billion in investments under its manage-ment was through secondary markets. Secondary markets allow Stafford to develop relationships with managers that often lead to future commitments.

Secondary markets are important because timberland returns in the United States have been compressed as the sector has matured and stump-age prices have remained low in the South over the past decade. Yet Stafford reasoned that investment returns could be improved by investing in a globally diversified portfolio and larger-scale assets, and by resetting the acquisition prices through secondary discounts. Indeed, Stafford's investment returns from secondary markets are generally higher than those from primary markets.

The development of secondary markets may help reduce some ineffi-ciency in the primary and mature timberland markets in the United States and elsewhere.

Open-End Funds Coming Back

In 1981, the First National Bank of Atlanta set up the first open-end timber-land fund in the United States. All other timberland funds since then have been closed end until recently. Since 2015, RMS, BTG Pactual, Hancock (for a timberland and agricultural fund), and Jamestown Invest have subsequently set up their own open-end funds. Interest has risen among TIMOs and their investors in open-end funds for several reasons.

First, the long slump in timber markets in the United States, especially in the South, has made some funds established between 2000 and 2007 unable to find willing buyers to whom they can sell their properties at appraised prices. Instead of selling at a loss and recognizing low realized returns, the owners of these funds have decided to ride it out and have extended their closing dates for a number of years. This in effect makes a closed-end fund "open" to some degree. Of course, some institutional investors (and/or TIMOs) decided to close their funds regardless, even if they had to offer a discount to reach a sale.

Second, more investors have come to realize that timberlands are really long-term investments, not just for ten to fifteen years. They also have some long-dated liabilities that need to be matched with real assets such as timberlands that produce short-term returns and long-term capital preservation. In other words, there are not many better alternatives to timberlands for utilizing their capital in the very long run.

One recent example of a transition from a closed- to an open-end fund is New Forests' Australia New Zealand Forest Fund, which originally had a ten-year closed-end structure and closed in October 2010 with AU$490 million (US$314 million). In May 2020, it was converted into a semipermanent vehicle with rolling twelve-year renewal periods. Only one investor decided to exit, and the units held by this investor were small and bought by New Forests using a small debt facility.[7]

Competition, Specialization, Consolidation, and Compressed Fees

TIMOs and REITs are always competing for investor capital and timberland investment opportunities. One executive in a publicly traded timberland REIT once told me that he could not understand why, with the well-managed and very liquid assets of timberland REITs, institutional investors would want to lock themselves in a TIMO-managed private timberland fund for a decade or longer. Perhaps different valuation metrics and the different types of investors that each investment vehicle attracts partially explain the coexistence of both investment vehicles. Empirical evidence presented in chapter 6 shows that these two investment vehicles have different risk-return profiles.

Competition among TIMOs has historically created specialization, consolidation, and even vertical integration. Many TIMOs still specialize in only a certain country or even only a certain region of a country. Some TIMOs focus on planted softwood forests, and others such as The Forestland Group concentrate on natural and hardwood forests. A few TIMOs with a conservation focus in their funds are good at conservation finance and therefore attract conservation-minded investors. TIMOs such as Hancock, Campbell Global, RMS, FIA, and GFP in the United States, New Forests in Australia, and BTG Pactual in Brazil have grown big in terms of assets under management and have invested on at least two continents. GFP now focuses exclusively on overseas markets. GFP has also pioneered investment in manufacturing strategies within its funds, having owned a paper mill in one fund and sawmills in several others. A few TIMOs such as Hancock, The Rohatyn Group, and Conservation Resource Partners are also investing in agricultural, infrastructural, and natural resource assets. As noted earlier, Stafford has focused on secondary markets.

Yet there are many firms—such as F&W Forestry (1.7 million acres) and Wagner Forest Management (1.3 million acres)—that have performed some TIMO activities but kept their traditional forestry consulting businesses. These traditional forest management and consulting firms also include American Forest Management (5.6 million acres), LandVest Timberlands (2.5 million acres), Forest Resource Consultants (2.2 million acres), and Prentiss & Carlisle (0.77 million acres). According to Forisk, these six firms collectively managed some 15 million acres of timberlands in the United States for individual and institutional investors or as subcontractors for TIMOs and timberland REITs as of May 2020.[8] Specialization and differentiation choices for investors help them better tailor their timberland strategies and diversify their exposures.

Competition among managers and newcomers to the TIMO business as well as more direct investments (see below) could result in lower management fees as well as mergers and acquisitions in the TIMO sphere. Some believe that if investment returns are curtailed because of a slump timber market, TIMOs will face pressure from their clients to cut fees. Others believe it is more likely that those managers will lose business or that investors will lose interest in the asset class. However, forces acting in the opposite direction, including more compliance costs and the low and even negative short-term interest rate environment, especially in Europe and Japan (which attracts investors investing in timberland), are also at work.

In the timberland REIT space, competition, the desire for scale, and other reasons have led to the disappearance of Crown Pacific Partners and U.S. Timberlands Company as well as the acquisition of Longview Fibre Company by a TIMO, Plum Creek Company by Weyerhaeuser, and Pope Resources by Rayonier, and the formation of PotlatchDeltic. The critical factor is the drive for scale. Without scale, a REIT risks becoming irrelevant or a target for a private equity takeover. As of today, Weyerhaeuser represents about 70 percent of the enterprise value in the publicly traded REIT space in the United States.

More Direct Investment in Timberland

Direct investors and the possible entry of more sophisticated managers who have not dealt with timberland before also compete with traditional TIMOs. Just like the Harvard Endowment once did, some large investors who wish to own real assets over the long term have started to insource investment teams or hire in-house managers who can help them achieve their goals and potentially reduce management fees. TIAA-CREF even bought out GreenWood Resources in 2012. Typically, these direct investors would like to have a large amount, say, $1 billion or more, to invest and see no reason to pay TIMO management fees, even if these are greatly reduced for large separate accounts. These direct investors in timberlands now include some US, Canadian, Australian, and New Zealand pension funds such as British Columbia Investment Management Corporation, Canada Pension Plan, Public Sector Pension Plan (Canadian), Ontario Teachers' Pension Plan, Alberta Investment Management, Australia Superannuation Fund, and New Zealand Superannuation Fund.

The entry of direct investors and more sophisticated managers also explains why some traditional forestry consulting firms, as noted above, have increasingly taken on some TIMO activities. For example, Yale University Endowment used Wagner Forest Management to manage its timber properties. This and other activities have gradually transformed Wagner Forest Management into a TIMO. As the line between forest management consulting firms and TIMOs becomes more blurred, it is possible that some management consulting firms could become more like TIMOs in the future.

Pushing Investment Frontiers

As noted earlier, Timberlink estimated that the total global institutional timberland investment through US-based TIMOs was $49 billion at the end of 2020.[9] According to Nareit, only 4 out of 192 publicly traded REITs in the United States were timberland REITs,[10] and the total enterprise value of these 4 firms was only about $39 billion, including some $7 billion in debt as of December 31, 2020. According to Stafford, global pension funds have grown to US$40 trillion in assets under management. Thus, after some thirty years of growth, timberlands are still a small sector and account for a very small share of total institutional investment worldwide.

Furthermore, some $31 billion, or nearly 64 percent, of the $49 Billion in institutional timberland investment through TIMOs took place in the continental United States. Oceania and South and Central America accounted for 25 percent and 10 percent of these investments, respectively, with Africa, Asia, and Europe all together accounting for the remaining 1 percent. As timberland markets in the United States mature and risk-adjusted returns decline over time, it is possible that future returns will not be as good as in the past. It is also possible that TIMOs and timberland REITs could go after some of the bigger nonindustrial private forest owners if they want to expand in the United States. In fact, some TIMOs have hired local forestry consultants to look for properties larger than five hundred acres and make an informal inquiry to see whether their owners are willing to sell. Should the inefficiency in nonindustrial timberland markets remain, these TIMOs could aggregate these lands and effectively deploy institutional capital. On the other hand, Atlanta-based Domain Timber Advisors closed 78 timberland sales to retail investors and other TIMOs with an average of 470 acres across the country in 2020.[11]

Better opportunities for institutional timberland investment in the next few decades may arise outside the United States. There are more (albeit riskier) investment-grade timberlands outside the United States, in areas where institutional timberland investment is currently concentrated. Oceania and South America could recover with strong demand in Asia. Europe could have new investment opportunities after timberland aggregation in western Europe and especially restitution of timberlands in central and eastern Europe, which is one of the last sources of "new" timberland available for institutional investment, although at a small scale. In Asia, where population pressure has often limited forest ownership to less than one hectare per family, local governments have aggregated small pieces of timberland to scale and rented them to institutional investors. The most difficult and potentially promising region of timberland investment is perhaps Africa, which is expected to account for more than 50 percent of the world's population growth by 2050. Some investors in Asia and Africa such as the Global Environment Fund have been rewarded with hefty returns, while others have experienced large losses. As long as there are higher returns that could compensate for the elevated level of risk, it is expected that some institutional timberland investors will continue to push the frontiers of investment on continents other than North America.

It is thus possible that a timberland index such as the NCREIF Timberland Index could emerge in other regions. Such an index would attract more institutional investors to invest in and expand the asset class sufficiently in these regions. Of course, the availability of infrastructure and forestry and management technical expertise in these regions would help bring capital there. Secure property rights and efficient governance structure are also critical, as many institutional investors are looking for investment opportunities with ESG (environmental sustainability, social responsibility, and responsible governance practices). At a forest investment forum organized by the World Bank and others in 2003, several TIMO managers strongly stressed the need for the World Bank and other institutions to create effective risk-reduction mechanisms and to help investors overcome high transaction costs in emerging markets.[12]

Impact Investing

Traditionally, many institutional investors have sought sustainable and value-aligned investments. For example, most institutional timberland investors have asked their managers to adhere to the sustainability principle by seeking forest certification, dialogues, and agreements with governments and conservation organizations regarding biodiversity and other environmental benefits on the land they own. They also invest in publicly traded timberland REITs that vow to stick to ESG goals. In this way, they align their values, embodied in their investments, with the broadly defined principle of sustainability.

Impact investments go one step further. Impact investments are defined as investments made in companies or organizations with the intent to contribute measurable positive social and environmental impacts, along with a financial return.[13] This definition encompasses three observable attributes of impact investors that can distinguish them from other investors: (1) *intention* to achieve social and environmental impacts, to be articulated before investment; (2) *contributions* to be made following a credible narrative or thesis and toward the achievement of the intended goals; and (3) *measurement* of delivery ex ante and ex post. Collectively, impact investments have the potential to make a difference in global development challenges such as poverty, inclusion, biodiversity, climate change, and land degradation.

International Finance Corporation estimates that as of 2018, impact investments were as high as $26 trillion—$21 trillion in publicly traded stocks and bonds, and $5 trillion in private markets involving private equity, nonsovereign private debt, and venture capital.

In the forestry sector, perhaps the Global Environment Fund (GEF) is the first that can be thought of as an impact investment because it meets all three criteria of impact investing. GEF covers other sectors as well but now has a forestry-specific fund (Africa Sustainable Forestry Fund, managed by Criterion Africa Partners). As noted earlier, unlike typical TIMOs, GEF raised money mostly from International Finance Corporation and other multilateral development finance agencies with explicit ESG goals. Thus, most of its investments have an underlying sustainability and/or a social development component, in addition to financial returns. GEF measures, quantifies, and reports its impacts mostly ex post.

To a large degree, all funds managed by Lyme Timber would qualify as impact investing because of its focus on "the acquisition and sustainable management of lands with unique conservation values," as would the Ecosystem Products Fund that New Forests raised in 2008. Another example of forestry-related impact investment is the Land Degradation Neutrality (LDN) Fund. Started by the United Nations Convention to Combat Desertification (UNCCD) secretariat and managed by Microva (an affiliate of Natixis Investment Managers in France), the LDN Fund is an impact investment fund investing in profit-generating sustainable land management and land restoration projects worldwide.

The LDN Fund provides long-term financing in both debt and equity for sustainable land use projects that prevent or reduce land degradation. Although the focus of this fund is sustainable land management, all of its first four projects so far are related to forestry, including a certified timber plantation with an agroforestry grower scheme in Ghana, agroforestry coffee cooperatives in Peru, a tree nut grower scheme in South Asia, and small-holder forestry in Kenya.[14]

Forest-Based Natural Climate Solutions

One of the biggest impact investment opportunities in forestry is related to nature-based solutions for climate change. The core idea for nature-based climate solutions is that plants are the only proven technology for actually removing CO_2 from the atmosphere, and trees are the preferred option because they not only take the CO_2 down but also store it for a long period. Furthermore, timber can be converted into long-lived building products that substitute for CO_2-intensive concrete and steel.

Forest-based natural climate solutions (FNCS) include afforestation, reforestation, and improved forest management as well as avoiding deforestation and forest degradation. While the latter are embodied in REDD+ programs and projects that are implemented mostly in forest-rich developing countries through international transfer payments,[15] the former are applicable in both developing and developed countries. Afforestation on antecedent land that was used for something other than forestry sequesters more CO_2 than other land uses. Rapid reforestation and restocking after natural events such as fires, windstorms, or insect attacks stores carbon quickly, as does purposeful timber harvesting. Improved forest management includes actions that increase the amount of carbon stored per unit area in existing forests by, for example, increasing stocking, extending rotation ages, or carrying out low-impact harvesting.

Recent peer-reviewed studies suggest that these approaches could make material contributions to reducing the amount of CO_2 in the atmosphere by up to 25 percent.[16] The "Mid-Century Strategy for Deep Decarbonization" produced in the waning years of the Obama administration suggests that FNCS would provide about 30 percent of the reductions required for the United States to meet its Paris Agreement targets.[17] More importantly, FNCS is a low-cost solution to sequestering carbon.[18]

There have been many efforts to establish carbon offset markets that allow legitimate carbon emitters to offset their emissions by obtaining equivalent reductions from sources of forest-based carbon sinks. Various voluntary carbon registration standards have emerged, and one market (the Chicago Climate Exchange) existed in the United States for nearly a decade before 2010. Currently the California carbon offset market registry accepts forest-based carbon projects and offers credits for carbon offsets if a forest property holds more carbon than the average of the US Forest Service "Forest Inventory and Analysis" region where the property is located. As noted earlier, this approach may invite adverse selection, as an investor could purchase a property with higher-than-regional average timber stocking, register and sell the "excess" carbon, and then resell the property. This may not be ideal from the perspective of combating climate change because the principle of additionality is violated, at least to some degree. Nonetheless, it established a market that allows forest investors and landowners to gauge the prevailing prices of carbon offset credits.

As mentioned in chapter 2, SilviaTerra is currently developing a natural capital exchange that allows the trading of offset credits generated from improved forest management activities such as postponing timber harvesting by private landowners.[19] Similarly, the American Forest Foundation and The Nature Conservancy, in collaboration with Verra (a nonprofit organization that oversees the Verified Carbon Standard, created The Family Forest

Carbon Program.[20] And Verra is planning to launch its own forest carbon exchange platform in 2021. Should these efforts succeed, it will be another way to establish a price for carbon offsets in forest management activities and provide an avenue for landowners to generate income from their forestlands.

It is foreseeable that with well-established and widely accepted carbon offset protocols and especially prices in the United States and elsewhere, institutional investors and their managers will start to actively seek investable opportunities for FNCS in the coming years. These opportunities could be projects that not only capture carbon and generate carbon offset income but also have other conservation cobenefits, local economic impact, commercial timber value, and attractive overall investment returns.

Blended Finance

Interestingly, the aforementioned LDN Fund uses a layered financial structure that blends public and private money together to achieve a broad public goal—land degradation neutrality—and to meet the demand of private investors for adequate returns. The essence of this kind of blended finance is to leverage public money to increase private sector investment in sustainable development.

This marriage of public and private finance starts with recognizing that public finance can cover only a fraction of the emerging financing needs for sustainable development even though it might play crucial role in creating the enabling conditions and infrastructure and in attracting private sector finance. The bulk of the money for sustainable development must come from private sources. In other words, there is a significant gap between available public finance and the scaling-up demand for sustainable development projects. Blended finance is a means to help close this gap.

In traditional timberland investing, TIMOs raise monies mainly from institutional investors and build up forest asset portfolios that generate risk-adjusted returns, as shown in part A of figure 8.1. The model has worked well in mature markets where industrial infrastructure is well developed. However, it has not been very successful in many developing countries because of much higher risks and shorter exit periods.

Blended finance is a new model and raises funds from different sources, as shown in part B of figure 8.1.[21] This new model starts with donors, government agencies, and communities with grants and seed money to establish a fund and conduct a feasibility study and other groundwork necessary for a sustainable development project. Then private and public impact investors

Figure 8.1. Timberland investment: Conventional and blended finance models

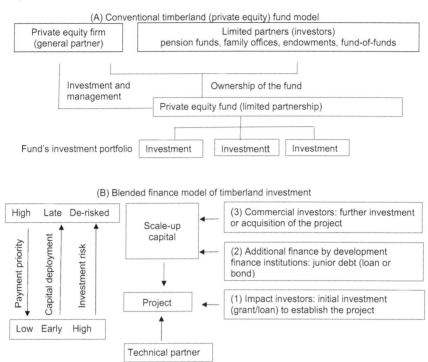

as well as development finance institutions come in as junior investors with loans and equity. Finally, commercial investors such as TIMOs and private banks come in as senior investors and contribute the bulk of the investment in the project.

In this blended finance mechanism, junior investors take a first-loss position in the fund, partially protecting senior investors. Junior investors are typically public organizations such as national development agencies, climate funds, or charity foundations. These investors target environmental and social development impacts first, before financial returns. They invest in a fund with blended finance because their junior participation catalyzes additional private investment in the senior layer, increasing the overall amount of investment in sustainable development.

On the other hand, senior investors are typically institutional investors such as pension funds, insurance companies, and private and development banks. These investors require market financial returns with a low risk profile, which are afforded by the layered structure of a blended fund. Usually, most senior investors are interested in additional positive development impacts but require these impacts to be in addition to an appropriate return-risk profile.

Interestingly, there are already several platforms and initiatives for sustainable financing and trade, sustainable forest management, and overall sustainable business development. Some are sector specific, others have a certain geographic focus, and still others have high-level sustainability strategies. Notable platforms and initiatives related to forestry and timberland investments include but are not limited to the (1) Climate Bonds Initiative; (2) World Business Council for Sustainable Development (WBCSD)—Forestry Solutions Group (FSG); (3) United Nations Forum on Forests Global Forest Financing Facilitation Network (GFFFN); and (4) Tropical Forest Alliance 2020 (TFA 2020).[22]

With the assistance of these platforms, along with the rise of impact investors, and especially the desire to close the financial gap between available public finance and the amount of investment needed to achieve global sustainability goals, it is foreseeable that public-private partnerships in the form of blended finance in timberland investment could flourish, especially in relation to forest and landscape restoration, biodiversity conservation, and forest-related climate change mitigation and adaptation.

Reintegration with the Forest Products Manufacturing Industry

When publicly traded forest products companies sold their timberlands in the United States in the 1990s and 2000s, private forest products firms—notably family-owned sawmills, plywood mills, and other manufacturing plants—did not follow. In fact, some private forest products firms have increased their timberland holdings, purchasing timberlands mostly from nonindustrial forest owners, to the extent that they have neglected the proper management of their existing timberlands.[23]

In Europe, two Nordic forestry companies sold their forests in the mid-2000s: Norske Skog and Stora Enso. This was not a planned capital allocation exercise but rather a reaction to necessity. Both Norske Skog and Stora Enso had made large unsuccessful acquisitions and had stretched their balance sheets. They sold their forests afterward. Stora Enso has largely reintegrated forestlands with forest products manufacturing since 2019. UPM-Kymmene Corporation (UPM) and Svenska Cellulosa AB (SCA), on the other hand, did not sell much of their forests at all, and they are two of the best-performing companies in terms of total returns. Some of the most successful cash-generating forestry companies in the world are Latin American companies, which are all vertically integrated as well and rely entirely on planted forests.[24] In fact, two large Chilean forest products companies—Compañía Manufacturera de Papeles y Cartones (CMPC) and Arauco (Celulosa Arauco y Constitución), each owning more than one million hectares of timberland in multiple Latin American countries—have rarely sold timberlands. In Brazil, Suzano (Suzano Papel e Celulose), which owns more than one million hectares of forestland (which foreign investors cannot), partners with TIMOs to develop pulp-wood plantations with supply agreements, much like the Travelers-Bowater arrangements in the United States many years ago.

Thus, empirically the backward integration from forest industry firms to timberlands still works, at least for privately held forest products firms in the United States and many forest products firms in Europe and Latin America. As noted earlier, publicly traded forest products firms in the United States sold their timberland partly because of accounting and taxation rules. Should these rules change, these companies could possibly buy some timberlands again. One factor that might lead to reintegration of timberland and manufacturing in the United States and elsewhere is the emergence of the International Financial Reporting Standards, which require mark-to-market accounting for such assets, avoiding the pitfalls of US GAAP.

Interestingly, timberland REITs in the United States have held forest products manufacturing facilities as part of the REIT or in a non-REIT subsidiary. One of the reasons that these REITs hold a manufacturing facility is that there is not an efficient timber market in places where they own timberlands. For example, Plum Creek once owned a sawmill in Montana because that was the only forest products manufacturing plant for timber produced from its forestlands. If it had not owned such a plant, Plum Creek would have been subject to the market power of a sole buyer of its timber or would have had stranded timberland assets if a second-party manufacturing plant closed its doors. Coincidentally, just as I completed the penultimate draft of this book, Timberland Investment Resources, a large TIMO in the United States, announced that it was going to build a state-of-the-art sawmill in Corinth, Mississippi, using capital provided by investors.[25]

Similarly, institutional investors who invested in South America, Asia, and Africa have found that forest products manufacturing markets are not well developed in many locations and that they may benefit from owning forest products manufacturing plants that process their timber. This forward integration from timberlands to forest products manufacturing requires more investment, but it is perhaps the only way to secure an adequate return for their timberland investment in some regions. In this case, without manufacturing to add value through processing, the economic value of standing timber is illusory.

Recalibrating Timberland Prices after COVID-19

When I was completing the first draft of this book, the COVID-19 global pandemic broke out. As a result, many economic activities were curtailed, some governments and businesses ceased operations completely for a while, and social distancing was introduced in the United States and other countries. A global recession was foreseen.

One of the timberland pricing models in the United States relates timberland prices positively to US GDP and negatively to long-term interest rates (ten-year US government bond rates).[26] If COVID-19 lingers for a couple of years and many governments and businesses continue to operate at a limited capacity, US GDP will fall, possibly dragging down US timberland prices. Fortunately, the Federal Reserve has lowered short-term interest rates effectively to zero and thus reduced long-term government rates. The net effect of these two events on timberland prices in the United States is thus uncertain.

On the other hand, because of increasing demand for hygiene products, including paper, and for wood products for do-it-yourself home improvement projects because of record-low interest rates as well as temporary shortages in labor and other production materials, forest products prices were at record levels in the summer and early fall of 2020 and early 2021. Even though high forest products prices did not trickle down to stumpage prices, at least in the US South,[27] those who could supply timber during this period might have secured high prices for their timber in other parts of the United States. Some timberland REITs have taken advantage of the situation and had a V-shaped recovery in their returns and stock prices.[28] In any event, all assets—including timberlands—could be recalibrated during and after the pandemic.

While the precise and long-term effect of COVID-19 on timberland investment is hard to predict, it is possible that deurbanization and working at home may gradually lift the demand for forest products and thus raise timber prices. It is also possible that some institutional timberland owners may need to sell their timberlands to raise capital to meet their obligations. Other investors may no longer believe that timber prices and returns in some regions are as attractive as before, and they may lose patience and sell their timberlands, seeking alternative investment opportunities elsewhere. Still others may see a great opportunity to get into timberland investment. It also is foreseeable that social distancing and the curtailment of travel could make TIMOs rely more on local consulting firms and high-tech methods of operation, such as using drones to measure inventory. Pushing the frontiers of timberland investment could also slow down for a while. If timberland prices worldwide are adjusted downward because of COVID-19 and if global economic growth returns to its historical average after the pandemic is over, we may find a great opportunity for new timberland investment in the next couple of years.

Notes

CHAPTER 1

1 W. Brad Smith et al., *Forest Resources of the United States, 2002*, General Technical Report NC-241 (St. Paul, MN: US Forest Service, North Central Research Station, 2004), 51–58.

2 Xing Sun and Daowei Zhang, "An Event Analysis of Industrial Timberland Sales on Shareholder Values of Major U.S. Forest Products Firms," *Forest Policy and Economics* 13 (2011): 396–401.

3 TIMO is an acronym coined by F. Christian Zinkhan in 1993. See Zinkhan, "Timberland Investment Management Organizations and Other Participants in Forest Asset Markets: A Survey," *Southern Journal of Applied Forestry* 17 (1993): 32–38. However, timberland in the United States is defined as forestland that is capable of producing 20 cubic feet of wood per year (or about 1.43 cubic meters per hectare per year) and has not been withdrawn from timber production for legal and administrative reasons. I continue to use TIMOs in this book with the understanding that TIMOs manage not only the narrowly and technically defined timberlands but also other forestlands. In other words, timberlands are the same as forestlands in this book.

4 Timberlink, "Timberland Assets under Management as of December 31, 2020," Atlanta, GA, www.timberlink.com.

5 Jim Hourdequin, "Some Thoughts on the Future of TIMOs and REITs" (paper presented at the State and Future of Forestry in the U.S. Meeting, May 29, 2013, Washington, DC).

6 David H. Newman and David N. Wear, "Production Economics of Private Forestry: A Comparison of Industrial and Nonindustrial Forest Owners," *American Journal of Agricultural Economics* 75 (1993): 674–84.

7 Michael Williams, *Americans and Their Forests: A Historical Geography* (New York: Cambridge University Press, 1989), 263–69.

8 The prevailing ad valorem property tax system in most timber-producing states in the country then was "merciless taxation," in which timber "was forced to pay annual revenues as though it were an annual crop." See Rupert B. Vance, *Human Geography of the South: A Study in Regional Resources and Human Adequacy*, 2nd ed. (Chapel Hill: University of North Carolina Press, 1935), as cited in William Boyd, *The Slain Wood: Papermaking and Its Environmental Consequences in the American South* (Baltimore: Johns Hopkins University Press, 2015). This tax system encouraged the premature cutting of timber while discouraging investment in forest conservation, making clear-cutting followed by abandonment more fiscally sensible than holding timber as a permanent investment. Fortunately, by the middle of the twentieth century, substantial progress had been made in rendering forest taxation more hospitable for timberland investment, including exemptions, modified assessments, yield taxes, and severance taxes.

9 International Paper Company, "Generation of Pride: A Centennial History of International Paper Company," reprinted as "A Short History of International Paper: Generations of Pride" in *Forest History Today*, 1998, 29–34.

10 International Paper Company, *1916 Annual Report*.

11 International Paper Company, *1919 Annual Report*, 8.

12 Note that trees and land are carried on the books of public companies at cost or fair market value, whichever is lower. As a result, a forest with a growing inventory is carried on the books at a lower valuation. The disparity between how timberland is valued by public market accounting and how it is valued as an investment asset is an important distinction and contributes to the ownership of most industrial timberlands by institutional investors who use the latter metric, and REITs that have single-level taxation.

13 International Paper Company, *1919 Annual Report*, 8.

14 International Paper Company, *1920 Annual Report*, 18-19.

15 International Paper Company, *1935 Annual Report*, 7.

16 "International Paper and Power," *Fortune* 16, no. 6 (1937): 229; Jack P. Oden, "Development of the Southern Pulp and Paper Industry, 1900-1970" (PhD diss., Mississippi State University, 1973), 578-79; Earl Porter and William Consoletti, *How Forestry Came to the Southeast: The Role of the Society of American Foresters* (Cenveo Publisher Services, 2015), 158-60; International Paper Company, *After Fifty Years*, 1948, 63-68.

17 I acknowledge Peter Mertz for pointing out the possible price-setting power of a forest products company on timberlands in the early twentieth century. I personally encountered a private landowner in his nineties in Alabama in 1998 who told me how open bidding on forestland sales worked in the South and how "reasonable" forestlands were priced in the 1930s.

18 International Paper Company, *After Fifty Years*, 63-68.

19 In July 2013, Weyerhaeuser Company bought 645,000 acres of timberland from Longview Timber LLC in Washington and Oregon. The purchase price for these timberlands was $2.65 billion, with an average price of $4,109 per acre. These timberlands could be much different from those it bought in 1900 in terms of location, species composition, maturity, and growth rate, as well as market conditions. Nonetheless, the size and general location of these two purchases are similar. If one assumes these timberlands are similar in terms of timber inventory, species composition, and maturity, the nominal rate of return for timberland investment during these 113 years was about 6 percent. Of course, this return accounts for only the capital appreciation component of timberland ownership. In other words, Weyerhaeuser Company also had income from timber harvesting, hunting leases, recreation, and mineral and development rights during these 113 years. If one assumes conservatively that the annual income generated was about the rate of timber growth, which was about 5 percent per annum in Washington and Oregon, the total return on timberland investment would be around 11 percent per year in nominal terms for Weyerhaeuser Company for much of the twentieth century. If one considers price appreciation and nontimber income as well, the total nominal rate of return during these 113 years would easily be 12-15 percent.

20 See "History," Weyerhaeuser, accessed April 15, 2016, http://www.weyerhaeuser. com/company/history/.

21 Weyerhaeuser Company, *1945 Annual Report*.

22 Weyerhaeuser Company, *1949 Annual Report*.

23 See "30 Years Change in Assets" subsection in Weyerhaeuser Company, *1945 Annual Report*.

24 Weyerhaeuser Company annual reports, various years.

25 Martin K. Perry, "Vertical Integration: Determinants and Effects," in *Handbook of Industrial Organization*, ed. Richard Schmalensee and Robert Willig (Amsterdam: North-Holland, 1989), 103-255. Empirically, Kurt Niquidet and Glen O'Kelly show that firms in New Zealand and Sweden decreased the proportion of fiber

sourced from a market with increasing fiber specificity, capital intensity, forest ownership concentration, and uncertainty. "Forest-Mill Integration: A Transaction Cost Perspective," *Forest Policy and Economics* 12 (2009): 207-12.

26 Norman D. Hungerford, "An Analysis of Forest Land Ownership Policies in the United States Pulp and Paper Industries" (PhD diss., State University College of Forestry at Syracuse University, 1968); Dennis W. Carlton, "Vertical Integration in Competitive Markets under Uncertainty," *Journal of Industrial Economics* 27, no. 3 (1979): 189-209; Jay O'Laughlin and Paul V. Ellefson, "Strategies for Corporate Timberland Ownership and Management," *Journal of Forestry* 12 (1982): 784-88; Paul V. Ellefson and Robert N. Stone, *U.S. Wood-Based Industry: Industrial Organization and Performance* (New York: Praeger, 1984); F. Christian Zinkhan et al., *Timberland Investments* (Portland, OR: Timber Press, 1992); Runsheng Yin, Thomas G. Harris, and Bob Izlar, "Why Forest Products Companies May Need to Hold Timberland," *Forest Products Journal* 50 (2000): 39-44.

27 James A. Rinehart, "Institutional Investment in U.S. Timberlands," *Forest Products Journal* 35, no. 5 (1985): 13-18.

28 Thomas P. Clephane and Jeanne Carroll, *Timber Ownership, Valuation, and Consumption Analysis for 97 Forest Products, Paper, and Diversified Companies* (New York: Morgan Stanley, August 25, 1982).

29 Also called systematic risk. In financial terms, a high level of risk means high volatility in earnings and stock prices.

30 Monopsonistic rents mean abnormal profits when one manufacturer controls its input prices. Monopsonistic rents arise when one buyer faces little competition from other buyers for an input and is able to set the prices for that input lower than they would be in a competitive market.

31 John R. Wilke, "How Driving Prices Lower Can Violate Antitrust Statutes: Monopsony Suits Mount as Companies Are Accused of Squeezing Suppliers," *Wall Street Journal*, January 27, 2004.

32 Brian Murray, "Oligopsony, Vertical Integration, and Output Substitution: Welfare Effects in the U.S. Pulpwood Markets," *Land Economics* 71, no. 2 (1995): 193-206; Brian Murray, "Measuring Oligopsony Power with Shadow Prices: U.S. Market for Pulpwood and Sawlogs," *Review of Economics and Statistics* 77 (1995): 486-98.

33 Boyd, *Slain Wood*, 58.

34 Ellefson and Stone, *U.S. Wood-Based Industry*; Rinehart, "Institutional Investment"; Yin, Harris, and Izlar, "Why Forest Products Companies"; Daowei Zhang and Peter H. Pearse, *Forest Economics* (Vancouver: University of British Columbia Press, 2011).

35 Yanshu Li and Daowei Zhang, "Industrial Timberland Ownership and Financial Performance of U.S. Forest Products Firms," *Forest Science* 60, no. 3 (2014): 569-78.

36 Yin, Harris, and Izlar, "Why Forest Products Companies"; Lars Lönnstedt, "Industrial Timberland Ownership in the USA: Arguments Based on Case Studies," *Silva Fennica* 41, no. 2 (2007): 379-91.

CHAPTER 2

1 Part of this section is drawn from Daowei Zhang and Richard Hall, "Timberland Asset Pricing in the United States," *Journal of Forest Economics* 35, no. 1 (2020): 43-67, http://dx.doi.org/10.1561/112.00000448. I acknowledge comments received from Clark S. Binkley and David Wear.

2 Martin Faustmann is credited with the first correct capital theory model, "On the Determination of Value Which Forest Land and Immature Stands Possess for

Forestry" (1849), in Institute Paper 42 (1968), translated by M. Gane, Commonwealth Forestry Institute, Oxford University. William F. Hyde was the first to incorporate silvicultural costs within the Faustmann formula. See William F. Hyde, *Timber Supply, Land Allocation, and Economic Efficiency* (Baltimore: Johns Hopkins University Press, 1980).

3   Richard Hartman, "The Harvesting Decision When a Standing Forest Has Value," *Economic Inquiry* 14 (1976): 52–68.

4   Sun Joseph Chang, "Determination of the Optimal Growing Stock and Cutting Cycle for an Uneven-Aged Stand," *Forest Science* 27, no. 4 (1981): 739–44.

5   Daowei Zhang, *The Softwood Lumber War: Politics, Economics, and the Long U.S.-Canadian Trade Dispute* (Washington, DC: Resources for the Future Press, 2007).

6   Noel Perceval Assogba and Daowei Zhang. "Conservation Reserve Program and Timber Prices" (working paper, Auburn University School of Forestry and Wildlife Sciences, 2020).

7   Olli Haltia, email message to author, July 22, 2020.

8   James M. Vardaman, *How to Make Money Growing Trees* (New York: John Wiley & Sons, 1989), 28–30.

9   The next few paragraphs are based on John M. Davis, interview with author, December 19, 2014.

10 Vardaman, *How to Make Money*, 36–38.

11 Empirically, Steven G. Burak used long-term corporate bond rates, and Daowei Zhang and Richard Hall used ten-year US government bond rates. Furthermore, Zhang and Hall found that their results were similar if they used corporate bond rates. See Steven G. Burak, "Selection of the Expected Rate of Return in the Valuation of Timberland: The Determinants of the Timberland Discount Rate" (PhD diss., Auburn University, 2001); Zhang and Hall, "Timberland Asset Pricing."

12 I acknowledge Richard Hall for pointing this out to me.

13 Daowei Zhang, "Costs of Delayed Reforestation and Failure to Reforest," *New Forests* 50 (2019): 57–70, doi:10.1007/s11056-018-9676-y.

14 Scott Jones, interview with author, November 26, 2014.

15 California Environmental Protection Agency Air Resources Board, *Compliance Offset Protocol: U.S. Forest Projects*, June 25, 2015, accessed March 31, 2020, https://ww3.arb.ca.gov/cc/capandtrade/protocols/usforest/forestprotocol2015.pdf; T. Ruseva et al., "Additionality and Permanence Standards in California's Forest Offset Protocol: A Review of Project and Program Level Implications," *Journal of Environmental Management* 198 (2017): 277–88.

16 NCX, https://www.ncx.com.

17 Vardaman, *How to Make Money*, 31.

18 John A. Holmes, ed., *The Dictionary of Forestry* (Bethesda, MD: Society of American Foresters, 1998); Ralph D. Nyland, *Silviculture: Concepts and Applications*, 2nd ed. (Long Grove, IL: Waveland Press, 2007).

CHAPTER 3

1   As late as two centuries ago, the majority of land on the earth was either under communal ownership or owned by emperors, tsars, monarchs, or churches. It was not until the fifteenth century when the price of wool was rising that manors in England started to enclose land and make it private to improve the profitability of grazing. This enclosure movement then spread all over England, and the concept of privately owned land started to take root. Private landownership was further seeded outside Europe when European colonists began to explore new

land, convert wilderness into property, and rent land to natives who had always used it freely.

Prior to this period, there was also feudal land ownership like serfdom and the peasant economy emerging in China and eastern Europe, respectively, in which a large amount of land was owned by a noble or landlord while the majority was still owned by emperors or tsars. However, such land ownership is not normally accepted as representative of the private ownership we know today, because of the lack of private property features, such as their transferability being limited only to inheritance or bestowal rather than business transactions. In fact, this kind of ownership was still attached to politics. By contrast, after the establishment of the private property idea in England, the law, economy, and politics were also changed to accommodate private landownership. See Andro Linklater, *Owning the Earth: The Transformation History of Land Ownership* (New York: Bloomsbury USA, 2013).

2   Resources for the Future, *Forest Credits in the United States: A Survey of Needs and Facilities* (Washington, DC: Resources for the Future, 1958). Until 1953, forestlands had been considered "unimproved property" and hence not acceptable as collateral. The amendment of section 24 of the Federal Reserve Act was sponsored primarily by Pacific Northwest banking and timber interests. It authorized banks and other financial institutions to make mortgage loans secured by first liens on forest tracts that were "properly managed in all respects."

3   Dale Morrison, interview with author, November 6, 2014. Clearly, access to capital was not a problem for large industrial firms that could borrow in national capital markets before the amendment of the Federal Reserve Act in 1953.

4   As cited in Boyd, *Slain Wood*; G. A. Fletcher, "Timberlands as Long-Term Loan Investment for Insurance Companies," *Forest Farmer* 26 (1967): 15. As for southern banks, the Citizens and Southern National Bank of Georgia (C&S) was one of the first and most active banks to get involved in timberland loans, often teaming up with life insurance companies such as Travelers. See C. M. Chapman, "A Private Banker Looks at Timber Loans," *Forest Farmer* 26 (1967): 16. C&S Bank of Georgia was the largest bank in the Southeast for much of the twentieth century. It later merged with other banks, including North Carolina National Bank (NCNB), to form the core of today's Bank of America, which continues to do mortgage lending business for timberland.

5   Letter from G. A. Fletcher, regional manager of agricultural loans, Travelers, to Guy Wesley, vice president of National Turpentine & Pulpwood Corporation, January 19, 1955. In author's files.

6   The Travelers story came from the author's interaction with Charlie F. Raper before his death, and interviews with Charles VanOver on March 12, 2012, and Dale Morrison on November 6, 2014.

7   From interviews with Charley Tarver, founder of Forest Investment Associates Inc., on December 12, 2011, and Dale Morrison on November 6, 2014.

8   ERISA of 1974 required asset diversification for a private company employee retirement account. All fifty US states followed with similar stipulations on their public pensions. These laws in effect put real estate, including timberlands, on the radar of institutional investors.

9   George Mason, "Role of Insurance Companies in Timberland Investment," in *Timberland Marketplace for Buyers, Sellers, and Investors and Their Advisors: Proceedings of 1984 Meetings* (Durham, NC: Center for Forestry Investment, School of Forestry and Environmental Studies, Duke University, March 1986), 163–67.

10   Charley Tarver, interview with author, February 20, 2015.

11 Ken Hines, interview with author, February 2, 2016.

12 Richard N. Smith, "A Proposal for CalPERS Separate Timber Account," memorandum to Richard P. Troy, January 28, 1985. As shown in this mero, Travelers had twenty foresters around the country at the time. In author's files.

13 Jake P. Petrosino, phone interview with author, March 12, 2015.

14 Claudia Roberts Pitas, "Timberland Investment by Pension Funds," in *Timberland Marketplace for Buyers, Sellers, and Investors and Their Advisors: Proceedings of 1984 Meetings* (Durham, NC: Center for Forestry Investment, School of Forestry and Environmental Studies, Duke University, March 1986), 176–78.

15 CalPERS, *Minutes of Investment Committee Meeting*, July 18, 1985.

16 Charles VanOver, interview with author, March 12, 2012.

17 Richard N. Smith, interview with author, January 9, 2015.

18 Wagner Southern was the predecessor of Forest Investment Associates Inc. It was set up primarily by Charley Tarver and Hank Swan of Wagner Woodlands and had the financial backing of a British investment bank between 1984 and 1985. See the next subsection.

19 Jake Petrosino, phone interview with author, March 12, 2015.

20 Smith, "Proposal for CalPERS"; Richard N. Smith, interview with author, January 9, 2015.

21 CalPERS, *Minutes of Investment Committee*, February 11, 1987.

22 CalPERS, *Minutes of Closed Session of Investment Committee*, June 16, 1987.

23 CalPERS, *Annual Financial Report of 1987*, 24.

24 Goldman Sachs, *Our Expanding Universe: The Case for Pension Fund Investment in Property* (New York: Goldman Sachs, 1983). This publication shows that the real annual rates of return for common stock, long-term government bonds, short-term government Treasury bills, and real property between 1970 and 1982 were 0.8%, -1.2%, 0.4%, and 4.5%, respectively. Similarly, an article by Preston E. Kirk shows that the annual rates of return between 1960 and 1984 for the S&P 500, corporate bonds, Treasury bills, and real estate were 3.5%, -0.1%, 1.1%, and 6.2%, respectively, while that for timberland was 6.4%. "A Tale of Tall Timber: Trying Times in the Tree Business Offer Investment Opportunity for Some," *Muse Air Monthly*, March 1985, 15–18. As we will see later, this lack of investment returns was also a reason for another pioneer, Eric Oddleifson, to start another TIMO, Boston Company Resource Investments, in 1982.

25 O. Medwin Welstad, interview with author, February 25, 2015.

26 Peggy A. Wong, *Forestry Fund Investment Study* (Marketing Research Department, US National Bank of Oregon, May 1982).

27 Robert M. Janzik, "Role of Pension Funds in Timberland Investment: Overview of Our Approach," in *Timberland Marketplace for Buyers, Sellers, and Investors and Their Advisors: Proceedings of 1984 Meetings* (Durham, NC: Center for Forestry Investment, School of Forestry and Environmental Studies, Duke University, March 1986), 111–13. The traits of timberland investment mentioned in three brochures by FNBA, Hancock for its ForesTree I Fund, and Oregon National Bank for its Collective Timberland Trust Fund were similar.

28 Cindi Clemmer, "Edwin Craig Wall Sr.: The Man behind the Empire," *Myrtle Beach Magazine*, Winter 1985, 52–57, 68.

29 Canal Industries Inc., *Celebrating 50 Years of Excellence in the Forest Products Industry: 1937–1987* (Conway, SC: Canal Industries Inc., 1987).

30 Charles Godfrey, phone interview with author, February 25, 2015.

31 Charles Dyer, phone interview with author, February 25, 2015.

32   From Sherry Smith, who pulled the filing documents for these companies from South Carolina's Secretary of State Office. In author's files.

33   Robert G. Chambers, interview with author, December 23, 2014.

34   "Frederick Wagner, N.H. Landowner Dies," Newspaper Archive of Journal Opinion, November 11, 1981, accessed May 1, 2020, http://jop.stparchive.com/Archive/JOP/JOP11111981P07.php.

35   Ken Super, interview with author, January 12, 2015; David Roby, phone interview with author, March 31, 2015.

36   Bob Berti, phone interviews with author, March 27 and 30, 2015.

37   Ken Super, interview with author, January 12, 2015.

38   Ed Hutcheson, interview with author, December 19, 2014.

39   Peter R. Stein, interview with author, January 12, 2015; David Roby, interview with author, March 31, 2015.

40   Terri Thompson, "Luring Pension Mangers into the Woods," *Business Week*, February 11, 1985, 89, 92.

41   The Campbell Group, *Business Plan Summary* (1984). This document also states: "Unique economic and tax considerations have created an environment in which timber as an investment has become a viable component of an investor's diversified portfolio via a partnership unit." It also said: "The business focus of The Campbell Group is to raise capital for the acquisition and management of timber resources. We intend to accomplish this goal by raising capital primarily from individual investors, pension plans and/or trusts in order to provide capital to the forest products industry that it will need in the 1980s."

42   Duncan Campbell, phone interview with author, March 16, 2015.

43   Duncan Campbell, "Foreign Capital in U.S. Timberland Investment: Who Invests, Why, and What the Futures May Bring," in *Timberland Marketplace for Buyers, Sellers, and Investors and Their Advisors: Proceedings of 1984 Meetings* (Durham, NC: Center for Forestry Investment, School of Forestry and Environmental Studies, Duke University, March 1986), 38–42. Although the printed version read "especially for international investors," after reading the whole article, I think this could be a typo and perhaps should be "especially for institutional investors."

44   The same proposition was independently discovered in a study by Pension Realty Advisors Inc., at the request of CalPERS. See Pitas, "Timberland Investment."

45   Clark S. Binkley, email message to author, November 5, 2020.

46   The Campbell Group, *Business Plan Summary*.

47   See figure 3.2 for an illustration.

48   Obtained via an email from Eva Greger, November 25, 2014.

49   It was unclear whether Eric Oddleifson and Ed Broom knew of the timberland funds in First National Bank of Atlanta at the time.

50   Eva Greger, interview with author, November 12, 2014; Ed Broom, phone interview with author, December 5, 2014.

51   Billy Humphries, email message to author, December 11, 2014.

52   This proposal would resurface again when the United Kingdom was under the Conservative Party leadership of John Major in the 1990s.

53   Alexander "Zan" Fell, phone interview with author, February 2, 2015.

54   Rosemary Howell, *The Fountain Forestry: The First Fifty Years* (Ilfracombe, United Kingdom: Arthur H. Stockwell Ltd., 2008).

55   Bob Flynn, email message to author, April 20, 2017. Bob pointed me to the website of Forestal Caja Bancaria, http://www.forestalbancaria.com.uy (accessed April 25, 2017).

56 Theodore E. Howard and Susan E. Lacy, "Forestry Limited Partnerships: Will Tax Reform Eliminate Benefits for Investors?," *Journal of Forestry* 84, no. 12 (1986): 39–43; Adrian M. Gilbert, "Investing in Timberlands—The Hutton Case," *The Consultant*, January 1984, 14–16.

CHAPTER 4

1   Jim Webb Jr., interviews with author, November 26, 2014, and March 22, 2015.
2   Scott Jones, phone interview with author, November 26, 2014.
3   Kirk, "Tale of Tall Timber"; Joe P. Mattey, *The Timber Bubble That Burst: Government Policy and the Bailout of 1984* (New York: Oxford University Press 1990).
4   Kirk, "Tale of Tall Timber."
5   "Investor: US & Major Takeovers," Sir James Goldsmith, accessed June 15, 2015, http://www.sirjamesgoldsmith.com/businessman/investor/us-major-takeovers/.
6   John M. Berry, "James River Builds Paper Empire," *Washington Post*, December 10, 1984.
7   As we will see later, some TIMOs are engaging in forward integration into logging and sawmills.
8   Gerald F. Davis and Tracy A. Thompson, "A Social Movement Perspective on Corporate Control," *Administrative Science Quarterly* 39 (1994): 141–73.
9   Ivan Fallon, *Billionaire: The Life and Times of Sir James Goldsmith* (Boston: Little, Brown, 1989), 359.
10 The book value of an asset is the value at which the asset is carried on a balance sheet and is calculated as the purchase price minus the accumulated depreciation.
11 Clephane and Carroll, *Timber Ownership*. Granted, the assumptions used in the Morgan Stanley study could be challenged. Nonetheless, this discrepancy is widely recognized by forest industry firms themselves. For example, the 1978 annual report of International Paper stated: "The demand for limited natural resources in recent years has increased the value of these lands substantially. *Current market value far exceeds the carrying value of the Company's land on the balance sheet*" (12) (emphasis added).
12 Hungerford, "Analysis of Forest Land Ownership," 121–22.
13 I appreciate Peter Mertz for emphasizing this point to me.
14 Kathleen K. Wiegner, "A Growing Investment?," *Forbes*, November 8, 1982, 52–53. It should be pointed out that the undervaluation of timberland in a company's book value is not unique, as book value cannot reflect current, true, or exact market values of mills or other assets, either. All fixed assets are reflected on the books at cost less depreciation/depletion. However, what makes timberland different from any other fixed assets such as a factory, equipment, or nonrenewable resources is that timber grows, and the gap between the current market value of timberland and its book value becomes much larger than that of other assets over time.
    It is also true that Wall Street values a company based not on its book assets but on its current earnings, and earnings potential—return on investment, return on capital, return on equity, cash flow, and cash available for distribution. Manage the asset to maximize earnings, and Wall Street will reflect that in share prices accordingly. Thus, in retrospect, if Crown Zellerbach had managed its timberlands to maximize earnings, and if its timberlands had been run as a separate accountable division, then Sir James would have been forced to pay some multiple of earnings to get his hands on them. In this regard, running their timberland

divisions as cost centers was the main issue that faced forest products companies in the 1980s, not the low book value of their timberlands, because the timberland assets were not earning. I appreciate Douglas Charles for emphasizing this point to me.

15 A white knight is a friendly investor that acquires a company's share with support from the company when it faces a hostile takeover.

16 A poison pill, or "shareholder rights plan," is a special dividend issued by the board of directors to existing stockholders to prevent a hostile takeover. Once triggered, the typical poison pill works by allowing existing shareholders to purchase stock at a deep discount, thus diluting the value of the stock and making a takeover more expensive for the hostile acquirer. Therefore, managers who succeed in getting a poison pill adopted need to have substantial discretion in governing the corporation. See Gerald F. Davis, "Agents without Principles? The Spread of the Poison Pill through the Intercorporate Network," *Administrative Science Quarterly* 36 (1991): 583–613.

17 Andrew A. Gunnoe, "Seeing the Forest from the Trees: Finance and Managerial Control in the U.S. Forest Products Industry, 1945–2008" (PhD diss., University of Tennessee, 2012).

18 A put option is an option contract giving the contract owner the right, but not the obligation, to sell a specified amount of an underlying security at a specified price within a specified time. In our case, Goldsmith had the right to sell the timberlands to Travelers for $250 million within a couple of years, if he wanted.

19 "Jimmy Goldsmith's US Bonanza," *Fortune*, October 17, 1983, 125–36. Interestingly, the 1980s buyouts by investors such as Sir James Goldsmith were portrayed in a 1987 film titled *Wall Street*. Michael Douglas won an Academy Award for best actor for the character Gordon Gekko in this film. The most iconic lines from his portrayal of Gordon Gekko come during a speech at a shareholders' meeting of Teldar Paper Company in which he complains about logging trucks and twenty-five vice presidents in the company. This pop culture portrayal lines up very closely with what I cover in this section. Also, the writer and director of *Wall Street*, Oliver Stone, has noted that one of the other characters in the film, Sir Lawrence Wildman, is based on Sir James Goldsmith.

20 "Investor: US & Major Takeovers," Sir James Goldsmith, accessed April 10, 2016, http://www.sirjamesgoldsmith.com/businessman/investor/us-major-takeovers/.

21 Clephane and Carroll, *Timber Ownership.*

22 Robert Chambers, phone interview with author, December 12, 2014. Chambers was cofounder of Timberland Investment Services, which became Timbervest LLC.

23 James Bearrows, "Other Financial Alternatives concerning Timberland Investment: Limited Partnership," in *Timberland Marketplace for Buyers, Sellers, and Investors and Their Advisors: Proceedings of 1984 Meetings* (Durham, NC: Center for Forestry Investment, School of Forestry and Environmental Studies, Duke University, March 1986), 17–21.

24 "A Poison Pill Has the Pros Choking," *Business Week*, February 11, 1985, 96.

25 Howard and Lacy, "Forestry Limited Partnerships."

26 Gunnoe, "Seeing the Forest."

27 Mike Clutter et al., *Strategic Factors Driving Timberland Ownership Changes in the U.S. South*, 2005, accessed July 8, 2016, https://www.iatp.org/sites/default/files/181_2_78129.pdf.

28 James Rinehart, *U.S. Timberland Post-recession: Is It the Same Asset?*, R&A Investment Forestry, April 2010, accessed December 31, 2016, http://

investmentforestry.com/resources/1%20-%20Post-Recession%20Timberland. PDF.

29 I thank Clark S. Binkley for pointing this out to me.

30 Lehman Brothers, *Timber 101: Valuing Timberland* (New York: Lehman Brothers, 2006).

31 Li and Zhang, "Industrial Timberland Ownership."

32 (100 × 0.21 + 100 × 0.21 × 0.20) / 100 = 36.8.

33 On the other hand, the long timber production process does not imply that all timberland investments must be long term, as there are cases in which even institutionally owned timberlands have changed hands within a couple of years. Also, there is a misconception that final harvest should coincide with the sale of the land. This is not required because a competitive market often attributes the right economic value of the standing timber on the land when it is sold. Yet one of the main hypotheses of this book is that timber and timberland markets are not perfectly competitive as defined in economics textbooks, allowing smart investors to arbitrage and profit from timberland investment.

34 29 U.S. Code § 1104.

35 Also called "prudent person rule" or "prudent investor rule," this rule originated from a legal case (*Harvard College v. Amory*) in which Massachusetts justice Samuel Putnam directed trustees who govern trust investments "to observe how men of prudence, discretion and intelligence manage their own affairs, not in regard to speculation, but in regard to the permanent disposition of their funds, considering the probable income, as well as the probable safety of the capital to be invested."

36 R. Smith and K. H. Bacon, "Boom in Tax Shelters Artificially Lifts Prices of Much Real Estate," *Wall Street Journal*, December 27, 1983.

37 CalPERS, *Investment Committee Meeting Minutes: John Hancock Timberland Portfolio Review*, September 20, 1988.

38 Richard N. Smith, interview with author, January 9, 2015. Smith acknowledged that this recommendation was also made by Ed Givhan of Resource Management Service.

39 The US government declared on June 22, 1990, that the northern spotted owl was threatened and would be placed under the protection of the federal Endangered Species Act, effective July 23, 1990. Hongshu Guan and Ian A. Munn, "Harvest Restrictions: An Analysis of New Capital Expenditures in the Pacific Northwest and South," *Journal of Forestry* 98, no. 4 (2000): 11–16.

40 Ryan Dezember, "Tree Gluts Uproot Southern Investors," *Wall Street Journal*, October 10, 2018.

41 Richard N. Smith, "Role of Insurance Companies in Timberland Investment," in *Timberland Marketplace for Buyers, Sellers, and Investors and Their Advisors: Proceedings of 1984 Meetings* (Durham, NC: Center for Forestry Investment, School of Forestry and Environmental Studies, Duke University, March 1986), 203–7.

42 Kirk, "Tale of Tall Timber."

43 The coefficient of variation represents the ratio of the standard deviation to the mean (of the returns of an investment). In finance, the coefficient of variation allows investors to determine how much volatility, or risk, is assumed compared to the amount of return expected from an investment. The lower the coefficient of variation, the better the risk/return trade-off. The Sharpe ratio is the average return earned in excess of the risk-free rate per unit of *volatility* or total risk, where the risk-free rate is measured by the yield or rate of the US Treasury, and volatility (or total risk) is represented by the standard deviation of an investment.

In practice, the risk-free rate may be represented by one-month, three-month, one-year, or even two-year US Treasury yields. In this book, the risk-free rate is represented by the three-month US Treasury yield unless noted otherwise. The higher the Sharpe ratio, the better.

44 Kirk, "Tale of Tall Timber."

45 See Zhang and Pearse, *Forest Economics*. The cost approach generates an estimate of the cost of replacing an asset (a forest in our case). For a premerchantable timber stand, this means compounding all the costs or expenses incurred in afforestation/reforestation and stand treatments to the age of the stand using an appropriate interest rate. The income approach produces an estimated value based on an asset's projected income. Under this approach, the value of a premerchantable timber stand is equal to the timber stand's projected revenues discounted to the stand's age. This estimated value of timber—using either approach, plus the value of bare land (dirt), which is based on comparable sales— is the estimated value of the forestland (forest and land). Yet the cost approach was mostly used by the forest industry in land purchases in the 1970s and 1980s. See William Sizemore, "Valuation and Appraisal," in *Timberland Marketplace for Buyers, Sellers, and Investors and Their Advisors: Proceedings of 1984 Meetings* (Durham, NC: Center for Forestry Investment, School of Forestry and Environmental Studies, Duke University, March 1986), 193–202.

46 Zinkhan et al., *Timberland Investments*.

47 Fred Tuemmler, phone interview with author, February 11, 2016.

48 CalPERS, *Hancock Timber Investment Information to Members of the Investment Committee* (Sacramento, CA: CalPERS Investment Office, January 19, 1988).

49 Emmett F. Thompson, interview with author, June 30, 2017.

50 David Newman, conversation with author, November 5, 2016.

51 Conversely, it is also clear now that the $7–10 billion or so invested in US timberland between 2005 and 2008 has performed very poorly. Depending on whether acquisitions during this period employed leverage, they have probably averaged between a negative (with leverage) and single-digit rate of return over twelve to fourteen years.

52 This strategy of harvest timing does not always work. The worst case is that everyone adopts the same strategy by exploring the option value of harvesting, which could lead to a prolonged slump in timber prices.

53 Richard N. Smith, interview with author, January 9, 2015.

54 Douglas Charles, phone interview with author, December 17, 2014.

CHAPTER 5

1 W. L. Mills and William L. Hoover, "Investment in Forest Land: Aspects of Risk and Diversification," *Land Economics* 58 (February 1982): 33–51; Clair H. Redmond and Frederick W. Cubbage, "Risk and Returns from Timber Investments," *Land Economics* 64 (November 1988): 325–37.

2 Robert Conroy and Mike Miles, "Commercial Forestland in the Pension Portfolio: The Biological Beta," *Financial Analysts Journal* 45 (September/October 1989): 46–54.

3 William F. Sharpe, "Capital Asset Prices: A Theory of Market Equilibrium under Conditions of Risk," *Journal of Finance* 19 (1964): 425–42; John Lintner, "The Valuation of Risk Assets and the Selection of Risky Investments in Stock Portfolios and Capital Budgets," *Review of Economics and Statistics* 47 (1965): 13–37.

4   Michael J. Jenson, "Risk, the Pricing of Capital Assets and the Evaluation of Investment in Stock Portfolios," *Journal of Business* 42, no. 2 (1969): 167–247.

5   Changyou Sun and Daowei Zhang, "Assessing the Financial Performance of Forestry-Related Investment Vehicles: Capital Asset Pricing Model vs. Arbitrage Pricing Theory," *American Journal of Agricultural Economics* 83, no. 3 (2001): 617–28.

6   James A. Rinehart and Paul S. Saint-Pierre, *Timberland: An Industry, Investment, and Business Overview*, 2nd ed., prepared by Pension Realty Advisors Inc. under a contract funded by the Hancock Timber Resource Group (Boston: Hancock Timber Resource Group, 1991).

7   One of the more recent titles in the gray literature is Chung-Hong Fu, *Timberland Investments: A Primer* (Brookline, MA: Timberland Investment Resources, 2014).

8   Douglas Charles, phone interview with author, December 17, 2014.

9   On September 17, 2015, the Securities and Exchange Commission (SEC) ruled that Timbervest "committed fraud by making material misrepresentations and omissions, and its principals aided, abetted, and caused the violations." The SEC barred the firm's principals from the investment advisory business for five years and ordered that they pay $403,500 plus $181,405 in interest, pending appeals. See David Allison, "SEC orders Atlanta's Timbervest to Pay $585,000," *Atlanta Business Chronicle,* September 22, 2015, accessed November 16, 2015, http://www.bizjournals.com/atlanta/news/2015/09/22/sec-orders-atlantas-timbervest-to-pay-585-000.html.

10 Hancock Timber Resource Group, *A Timber Investment Program for the California Public Employees Retirement System (CalPERS): Proposal—Timberlands RFP #85-31* (Boston, 1986).

11 One can imagine how big these incentive fees were, assuming that HTRG took 20 percent if the rate of return was greater than 8 percent in real terms. The average annual total return of the NCREIF Timberland Index in the Pacific Northwest—nearly all managed by HTRG—was 35.7 percent between 1987 and 1995. The annual average asset value in the region in the same period was $539 million. See Hancock Timber Resource Group, *Research Notes: The NCREIF Timberland Index* (Boston, 1996), 5–8.

12 This and the next paragraph are from L. Richard Doelling, phone interview with author, July 14, 2016.

13 In this situation, the appraisal may be too high. Sometimes the cash flows modeled in an appraisal bear little resemblance to the cash flows that managers are generating. This implies that either the appraisal is too optimistic or the managers are not optimizing the investments.

14 Jim Webb Jr., interview with author, March 25, 3015.

15 Daowei Zhang, Brett Butler, and Rao Nagubadi, "Institutional Timberland Ownership in the U.S. South: Magnitude, Location, Dynamics, and Management," *Journal of Forestry* 110, no. 7 (2012): 355–61.

16 Zinkhan, "Timberland Investment Management."

17 Ian A. Munn and W. Rhett Rogers, "Forest Management Intensity: A Comparison of Timber Investment Management Organizations and Industrial Landowners," *Southern Journal of Applied Forestry* 27, no. 2 (2003): 83–91; Kathryn G. Arano and Ian A. Munn, "Evaluating Forest Management Intensity: A Comparison among Major Forest Landowner Types," *Forest Policy and Economics* 9, no. 3 (2006): 237–48; Xing Sun, Daowei Zhang, and Brett Butler, "Timberland Ownerships and Reforestation in the Southern United States," *Forest Science* 61, no. 2 (2015): 336–43.

18 Peter R. Stein, "Trends in Forestland Ownership and Conservation," *Forest History Today*, Spring/Fall 2011, 83–86.

19 Charles Collins, "Examples of Recent Timberland Transactions," in *Industrial Timberland Divestitures and Investments: Opportunities and Challenges in Forestland Conservation*, ed. Nadine E. Block and V. Alaric Sample (Washington, DC: Pinchot Institute, 2001), 29–31.

20 See "International Paper Agrees to Sell 5.1 Million Acres of U.S. Forestland for Approximately $6.1 Billion," International Paper, accessed August 22, 2017, http://investor.internationalpaper.com/news-releases/Press-R/2006/International-Paper-Agrees-to-Sell-51-Million-Acres-of-US-Forestland-For-Approximately-61-Billion/default.aspx.

21 Stein, "Trends in Forestland Ownership."

22 These data are available at "Compliance Offset Program," California Air Resources Board, under "ARB Offset Credit Issuance," https://ww2.arb.ca.gov/our-work/programs/compliance-offset-program. Adverse selection in this case means that offset sales may be particularly attractive to firms whose true baselines are lower than the regulators' estimates and thus receive "windfall" profits by signing on to this program.

23 Cambridge Associates, *Timber Impact Investment Benchmark Preview, A Study for Global Impact Investment Network* (Boston: Cambridge Associates, 2017); Global Impact Investing Network, "GIIN Perspectives: Evidence on the Financial Performance of Impact Investments," accessed December 16, 2020, https://thegiin.org/assets/2017_GIIN_FinancialPerformanceImpactInvestments_Web.pdf.

24 See "International Paper Agrees to Sell 5.1 Million Acres of U.S. Forestland for Approximately $6.1 Billion," International Paper, accessed August 22, 2017, http://investor.internationalpaper.com/news-releases/Press-R/2006/International-Paper-Agrees-to-Sell-51-Million-Acres-of-US-Forestland-For-Approximately-61-Billion/default.aspx.

25 Brooks Mendell, "Wood Supply Agreements, Part I: Pricing Mechanisms and Market Analyses," *Forisk Blog*, May 9, 2013, accessed March 31, 2017, http://forisk.com/blog/2013/05/09/wood-supply-agreements-part-i-pricing-mechanisms-and-market-analyses/. For a theoretical analysis of timber supply contracts, see Runsheng Yin and Bob Izlar, "Supply Contract and Portfolio Insurance," *Journal of Forestry* 99, no. 5 (2001): 39–44.

CHAPTER 6

1 F. Christian Zinkhan, "The Stock Market's Reaction to Timberland Ownership Restructuring Announcements: A Note," *Forest Science* 34 (1988): 815–19. Similarly, several studies show that conversion from vertically integrated forest products firms to timberland REITs had positive impacts. See Brooks Mendell, Neena Mishra, and Tymur Sydor, "Investor Responses to Timberlands Structured as Real Estate Investment Trusts," *Journal of Forestry* 106 (2008): 277–80; Sun and Zhang, "Event Analysis"; Changyou Sun, Mohammad M. Rahman, and Ian A. Munn, "Adjustment of Stock Price and Volatility to Changes in Industrial Timberland Ownership," *Forest Policy and Economics* 26 (2013): 91–101.

2 Plum Creek, *Gratitude, Stewardship, Opportunity: 25 Years of Plum Creek* (Seattle: Plum Creek Timber Co., 2014).

3 "Crown Pacific Declares Bankruptcy," *The Forestry Source*, August 2003, 6.

4 Timber Growth Corporation, *Amendment No. 1 to Form S-11 Registration Statement under the Security Act of 1933* (Washington DC: Securities and Exchange Commission, April 29, 1998).

5 Rick Holley, phone interview with author, August 15, 2016.

6 Eleena de Lisser, "Georgia-Pacific Plans New Class of Stock Tied to Its Profitable Timber Business," *Wall Street Journal*, September 18, 1998.

7 This legislation and related tax rulings also opened the door for private REITs and, among other things, permitted non-US investors to avoid foreign real property tax when they disposed of their "U.S. real property interests" (USRPI) under the Foreign Investment in Real Property Tax Act (FIRPTA) of 1980. I acknowledge Clark S. Binkley for pointing this out to me.

8 See CatchMark, December 2, 2013, and July 8, 2018, http://www.catchmark.com/news/.

9 Lehman Brothers, *Timber 101*.

10 Vivian Marino, "For Some Investors, Money Grows on Trees," *New York Times*, May 27, 2007, accessed September 17, 2016, http://www.nytimes.com/2007/05/27/realestate/commercial/27sqft.html?_r=0.

11 Daowei Zhang and Shaun M. Tanger, "Is There a Connection between Campaign Contributions and Legislative Commitment? An Empirical Analysis on the Co-sponsorship Activity of the 2007 TREE Act," *Forest Policy and Economics* 85, no. 1 (2017): 85–94.

12 Sarah Borchersen-Kato, "Timberland REIT Potlatch to Acquire Deltic Timber in All-Stock Deal," Nareit, accessed June 6, 2020, https://www.reit.com/news/articles/timberland-reit-potlatch-acquire-deltic-timber-all-stock-deal.

13 On May 1, 2020, the stock price of Weyerhaeuser fell nearly 18 percent after it announced it would temporarily suspend dividend payments. See William White, "Weyerhaeuser Earnings: WY Stock Plunges 18% on Declining Demand," InvestorPlace, accessed May 20, 2020, https://investorplace.com/2020/05/weyerhaeuser-earnings-wy-stock-plunges-18-on-declining-demand/. One could argue that Weyerhaeuser had previously paid dividends that were too high a percentage of its free cash flow and came to realize that it could no longer do so in a global pandemic. This is not much different from the misjudgment of some TIMOs in 2010–2015 when they withheld timber from the markets because they expected timber prices to rise, which they did not do in the next five years.

14 Chung-Hong Fu, *A Look at the New NCREIF Timberland Fund and Separate Account Index*, Timberland Investment Resources, 2013, accessed April 20, 2020, https://1nzy1a2az6m43b6rbr2f9hib-wpengine.netdna-ssl.com/wp-content/uploads/2014/06/A-Look-at-the-NCREIF-Fund-Index-2013-07-11.pdf.

15 One may note the differences between these average compounded annual rates of return and the average annual rates of return reported in table 6.1. Students may want to resort to the definitions and calculations of geometrical and arithmetic means (of growth rates).

16 Eugene F. Fama and Kenneth R. French, "Common Risk Factors in the Returns on Stocks and Bonds," *Journal of Financial Economics* 33, no. 1 (1993): 3–56.

17 Wharton Research Data Services, accessed January 7, 2021, https://wrds-www.wharton.upenn.edu/.

18 This is confirmed by Bin Mei, who stated, "Timber REITs have some risk-reduction ability but no excess returns." See Bin Mei, "Timberland Investments in the United States: A Review and Prospects," *Forest Policy and Economics* 109 (2019): 1–7, doi.org/10.1016/j.forpol.2019.101998.

19 Bin Mei and Mike Clutter, "Evaluating the Financial Performance of Timberland Investments in the United States," *Forest Science* 56, no. 5 (2010): 421–28.

20 Sun and Zhang, "Assessing the Financial Performance."

21 F. Christian Zinkhan, "History of Modern Timberland Investments," in *Proceedings of Conference on International Forest Investment (London, UK)* (Athens, GA: Center for Forest Business, 2008).

22 Clark S. Binkley, *The Russell-NCREIF Timberland Index: A Review* (Vancouver, Canada: Faculty of Forestry, University of British Columbia, 1994); Mary Ellen Aronow, Clark S. Binkley, and Courtland L. Washburn, "Explaining Timberland Values in the United States," *Journal of Forestry* 102 (December 2004): 14–18.

23 Zinkhan et al., *Timberland Investments*; Sun and Zhang, "Assessing the Financial Performance"; Mei and Clutter, "Evaluating the Financial Performance."

CHAPTER 7

1 Richard W. Haynes, *Stumpage Prices, Volume Sold, and Volumes Harvested from the National Forests of the Pacific Northwest Region, 1984 to 1996*, General Technical Report PNW-GTR-423 (Portland, OR: US Forest Service Pacific Northwest Research Station, 1998); Guan and Munn, "Harvest Restrictions."

2 William T. Stanbury, *Environmental Groups and the International Conflict over the Forests of British Columbia, 1990 to 2000* (Vancouver, Canada: SFU-UBC Centre for the Study of Government and Business, 2000).

3 William H. Bradley, "Federal Income Tax—100 Years of Shaping Timberland Transactions: A Review of Past and Current Tax Policy and Some Predictions for the Future" (paper presented at Who Owns America's Forests Conference, Portland, OR, September 17–19, 2013).

4 Proskauer, "Significant Changes to U.S. Taxation of REITs and Investments by Non-U.S. Investors in Real Property under the PATH Act," February 2, 2016, accessed July 1, 2017, https://www.proskauer.com/alert/significant-changes-to-us-taxation-of-reits-and-investments-by-non-us-investors-in-real-property-under-the-path-act.

5 Peter H. Pearse, Daowei Zhang, and Janette Leitch, "Trends in Foreign Investment in Canada's Forest Industry," *Canadian Business Economics* 3, no. 3 (1995): 54–68.

6 Phillis Berman, "Viva St. Spotted Owl!," *Forbes*, February 12, 2006, 44–45; George Draffan, *The Global Timber Titans: Profile of Four U.S. Wood Products Corporations Driving the Globalization of the Industry*, June 1999, accessed August 9, 2020, https://bdgrdemocracy.files.wordpress.com/2011/12/globaltimbertitans.pdf.

7 Dennis Neilson, "Timberlands Ownerships in an International Context," *Western Forester*, March/April 2006, 6–7.

8 DANA Limited, *The New Zealand Forest Products Industry Database Review 2016*, 10th ed. (Rotorua, New Zealand: DANA Limited, 2016).

9 M. Clarke, "Devolving Forest Ownership through Privatization in New Zealand," *Unasylva* 50, no. 199 (1999): 35–44, http://www.fao.org/docrep/x3030e/x3030e0a.htm; New Zealand Institute of Economic Research, *Devolving Forest Ownership through Privatisation: Processes, Issues and Outcomes*, Working Paper 2000/3, May 2000, accessed June 30, 2016, http://nzier.org.nz/static/media/filer_public/c3/8b/c38bb726-0528-40f1-b9c3-4b8c15006cbe/wp2000-03_devolving_forest_ownership_through_privatisation.pdf.

10 Dennis Neilson of DANA Limited, phone conversation with author, February 3, 2017.

11 Bryce Heard, former CEO of Fletcher Challenge, phone conversation with author, February 5, 2017.

12 Tom Broom, phone interview with author, February 3, 2017.

13 HMC sold the entire property to three investors in 2013. This deal "has proved to be one of the most profitable plantation forests in the world for institutional investors (if not the most profitable) over the last decade." See DANA Limited, *New Zealand Forest Products*, 22–27.

14 "Weyerhaeuser Co.: Concern Seeks to Build Fund to Buy Foreign Timberlands," *Wall Street Journal*, April 14, 1995.

15 Peter Mertz, communication with author, August 17, 2020.

16 Reuters, "Finland's UPM to Invest $3 Billion to Uruguay," July 23, 2019, accessed November 24, 2019, https://uk.reuters.com/article/uk-upm-kymmene-oyj-results-uruguay/finlands-upm-to-invest-3-billion-in-uruguay-idUKKCN1UI0UV.

17 "Hancock Gets Clearance to Buy Carter Holt Harvey's Forestry Assets," *New Zealand Herald*, September 27, 2006, accessed August 14, 2020, https://www.nzherald.co.nz/carter-holt-harvey/news/article.cfm?o_id=36&objectid=10403251.

18 Campbell Global, "The Campbell Group Acquires South Australian Forestry Assets," October 17, 2012, accessed August 14, 2016, https://www.campbellglobal.com/about/press-release?mediaID=M127.

19 Campbell Global, "OneFortyOne Purchase of Nelson Forests Confirmed by Overseas Investment Office," August 20, 2018, accessed August 14, 2020, https://www.campbellglobal.com/about/press-release?mediaID=M141.

20 Neilson, "Timberlands Ownerships."

21 John Earhart, phone interview with author, April 5, 2017.

22 Michael McDonald and Irina Vilcu, "Harvard Exits Romanian Timber Woes, Selling Forests to Ikea," Bloomberg, July 23, 2015, accessed August 1, 2020, https://www.bloomberg.com/news/articles/2015-07-23/harvard-exits-romanian-timber-woes-selling-forests-to-ikea.

23 Juliet Chung and Dawn Lim, "Harvard Endowment Chief Pushed for Steeper Devaluation of Assets," *Wall Street Journal*, December 15, 2017, accessed August 1, 2020, https://www.wsj.com/articles/harvard-endowment-chief-pushed-for-steeper-devaluation-of-assets-1513252800.

24 New Forests, accessed July 3, 2020, https://www.newforests.com.au/.

25 "Stora Enso Finalises the Restructuring of Bergvik Skog's Forest Holdings," May 31, 2019, accessed August 14, 2019, https://www.storaenso.com/en/newsroom/regulatory-and-investor-releases/2019/5/stora-enso-finalises-the-restructuring-of-bergvik-skogs-forests-holdings.

26 James Ranaivoson and Enrico Canu, "EIB's Approach towards Forestry Funds" (presentation at roundtable meeting, Plantation Forestry in Developing Countries: Enhancing Innovation, Tackling Complexity, Luxembourg, May 22, 2012).

27 International Woodland Company, *25 Year Jubilee: 1991–2016*, Frederiksberg C, Denmark.

28 International Woodland Company, *25 Year Jubilee*; IWC, email message to author, August 2020.

29 "News and Insights," Stafford Capital Partners, accessed March 15, 2017, http://www.staffordcp.com/2016.

30 Marian Drăgoi and Veronica Toza, "Did Forestland Restitution Facilitate Institutional Amnesia? Some Evidence from Romanian Forest Policy," *Land* 8 (2019): 99, doi:10.3390/land8060099.

31 Marian Chiriac, "Much of Romania's Forest Restitution Deemed Illegal," *BalkanInsight*, October 17, 2014, accessed September 10, 2019, https://

balkaninsight.com/2014/10/17/illegalities-hamper-romania-s-forest-restitution-process/.

32 Mads Asprem, phone conversation with author, August 3, 2020.

33 African Development Bank, *Towards Large-Scale Commercial Investment in African Forestry* (study for Climate Investment Funds Evaluation and Learning Initiative, 2019).

34 Clark S. Binkley et al., *Pension-Fund Investment in Forestry* (Washington, DC: World Bank, 2021).

35 African Development Bank, *Unlocking Capital Flows for Forest Sector Development in Africa* (AfDB CIF Knowledge Series, July 2017).

36 Phaunos Timber Fund, accessed February 20, 2017, http://www.phaunostimber.com/.

37 James Carthew, "Cambium Global Timberland IT: Why Money Doesn't Grow on Trees," Citywire Funds Insider, June 14, 2011, accessed February 20, 2017, http://citywire.co.uk/money/cambium-global-timberland-it-why-money-doesn-t-grow-on-trees/a594817.

38 Conner Bailey et al., "Taking Goldschmidt to the Woods: Timberland Ownership and Quality of Life in Alabama," *Rural Sociology*, 2020, doi.org/10.1111/ruso.12344. Similarly, Madeleine Fairbairn has called attention to possible social and ecological issues related to institutional farmland investment. See Madeline Fairbairn, *Fields of Gold: Financing the Global Land Rush* (Ithaca, NY: Cornell University Press, 2020).

CHAPTER 8

1 See also Clark S. Binkley, "Long-Run Timber Supply: Price Elasticity, Inventory Elasticity and the Capital-Output Ratio," *Natural Resource Modelling* 7 (1994): 163–81.

2 Linda Wang, *Timber REITs and Taxation*, General Technical Report (US Forest Service, August 2011), accessed May 1, 2014, http://www.fs.fed.us/spf/coop/library/timber_reits_report.pdf.

3 Timberlink, *Timberland Assets*.

4 Dezember, "Tree Gluts."

5 Rayonier, "Rayonier Pacific Northwest Portfolio Repositioning Supplemental Materials," May 2016, accessed June 30, 2016, https://ir.rayonier.com/static-files/055f4b41-445c-4de6-b5fa-7004c29c7a97/.

6 Stephen Addicott, "Stafford's Secondary Approach to Timberland Investment" (presentation at RISI Timberland Investment Conference, London, UK, May 8, 2019).

7 See Daniel Kemp, "New Forests Converts ANZ Forest Fund to Semi-permanent Structure on 12-Year Terms," Agri Investor, May 7, 2020, accessed June 26, 2020, https://www.agriinvestor.com/new-forests-converts-anz-forest-fund-to-semi-permanent-structure-on-12-year-terms/.

8 Data provided by Brooks Mendell, August 17, 2020. The full list of forestland owners and managers can be found at "Forisk Timberland Owner List," Forisk Consulting, 2020, https://forisk.com/product/forisk-timberland-owner-list/.

9 Timberlink, *Timberland Assets*.

10 Nareit, accessed January 15, 2021, https://www.reit.com/.

11 Businesswire, accessed February 20, 2021, https://www.businesswire.com/news/home/20210125005060/en/Domain-Timber-Advisors-Achieves-Record-Year-in-Land-Sales

12 World Bank, *Report of the Forest Investment Forum: Investment Opportunities and Constraints* (Washington, DC: World Bank, 2004), accessed June 5, 2020, http://documents.worldbank.org/curated/en/143121468135583949/Report-of-the-Forest-Investment-Forum-investment-opportunities-and-constraints.

13 International Finance Corporation, *Creating Impact: The Promise of Impact Investing*, Washington, DC, April 2019, accessed May 25, 2020, https://www.ifc.org/wps/wcm/connect/66e30dce-0cdd-4490-93e4-d5f895c5e3fc/The-Promise-of-Impact-Investing.pdf?MOD=AJPERES.

14 It is estimated that two billion hectares of land are degraded worldwide, and another twelve million hectares of productive land are degraded every year. Land degradation threatens the future sustainability of the earth, with severe impacts on food security, livelihoods, climate change, biodiversity, and other ecosystem services. Sustainable land management practices such as landscape restoration and agroforestry offer great opportunities to reduce and reverse land degradation and produce significant environmental and social benefits. UNCCD defines land degradation neutrality as "a state whereby the amount and quality of land resources necessary to support ecosystem functions and services and enhance food security remains stable or increases." See Mirova, *Land Degradation Neutrality (LDN) Fund: An Innovative Impact Investment Fund for Sustainable Land Use, with a Linked TA Facility* (Paris, 2019).

15 REDD+ stands for reducing emissions from deforestation and forest degradation in developing countries, and the role of conservation, sustainable management of forests, and enhancement of forest carbon stocks in developing countries.

16 J.-F. Bastin et al., "The Global Tree Restoration Potential," *Science* 365, no. 6448 (2019): 76–79; J. Busch et al., "Potential for Low-Cost Carbon Dioxide Removal through Tropical Reforestation," *Nature Climate Change* 9, no. 6 (2019): 463.

17 White House, *United States Mid-Century Strategy for Deep Decarbonization* (Washington, DC: White House, 2016).

18 B. W. Griscom et al., "Natural Climate Solutions," *Proceedings of the National Academy of Sciences* 114, no. 44 (2017): 11645–50.

19 For more information see NCX, https://www.ncx.com/.

20 American Forest Foundation, accessed March 26, 2021, https://www.forestfoundation.org/family-forest-carbon-program.

21 Other examples of TIMOs and NGOs that use such fund structures include Finance in Motion, Moringa, Criterion Africa Partners, 12Tree Finance, International Woodland Company, and The Nature Conservancy.

22 Petri Lehtonen, *Forest Finance Hub Led by FAO? A Concept Note* (Rome: Food and Agriculture Organization of the United Nations, 2019). The Climate Bonds Initiative is an international organization working solely to mobilize capital for climate change solutions. Its strategy is to develop a large and liquid Green and Climate Bond Market that helps drive down the cost of capital for climate projects in developed and emerging markets and facilitate a rapid transition to a low-carbon and climate-resilient economy. The initiative has developed specific bond criteria for different sectors, including forestry.

WBCSD-FSG aims to bring more of the world's forests under sustainable management, expand markets for responsible forest products, and ensure growth and vitality of forest resources. FSG reports its members' (forest companies') progress in sustainability performance.

The UN Forum on Forests (UNFF) carries out its forest financing activities through the Global Forest Financing Facilitation Network (GFFFN). The functions of the GFFFN are as follows:

- promote the design of national forest financing strategies to mobilize resources for sustainable forest management;
- facilitate access to financing resources of all sources, including the Global Environment Facility and the Green Climate Fund;
- serve as a clearinghouse for existing, new, and emerging financing opportunities and as a tool for sharing lessons learned from successful projects; and
- contribute to the achievement of global forest goals and targets as well as priorities contained in the forum's Quadrennial Program of Work.

The Tropical Forest Alliance 2020 is a global public-private partnership in which partners take voluntary action to reduce the tropical deforestation associated with the sourcing of commodities such as palm oil, soy, beef, and paper and pulp.

23 Zhang, "Costs of Delayed Reforestation."

24 Mads Asprem, email message to author, July 22, 2020.

25 Y'all Politics, "Mission Forest Products Locating Sawmill in Corinth, Creating Approximately 130 Jobs," September 25, 2020, accessed September 30, 2020, https://yallpolitics.com/2020/09/25/mission-forest-products-locating-sawmill-in-corinth-creating-approximately-130-jobs/.

26 Zhang and Hall, "Timberland Asset Pricing."

27 Adam Maggard and Daowei Zhang, "Why Did Stumpage Prices Remain Low When Lumber Prices Soared to Historic Highs?," *Alabama TREASURE Forest Association Newsletter* 6 , no. 3 (December 2020): 1.

28 Kevin Sun, "Timber REITs Are Having a Mighty Moment," *The Real Deal*, August 20, 2020, accessed August 28, 2020, https://therealdeal.com/2020/08/20/timber-reits-are-having-a-mighty-moment/.

References

Adams, Darius M. 1983. "An Approach to Estimating Demand for National Forest Timber." *Forest Science* 29 (2): 289–300.

Adams, Darius M., and Richard W. Haynes. 1980. *The 1980 Softwood Timber Assessment Market Model: The Structure, Projections and Policy Simulations.* Forest Science Monograph 22.

———. 1989. "A Model of National Forest Timber Supply and Stumpage Markets in the Western United States." *Forest Science* 35 (2): 401–24.

Adams, Darius M., Richard W. Haynes, George F. Dutrow, Richard L. Barbier, and Joseph M. Vasievich. 1982. "Private Investment in Forest Management and the Long-Term Supply of Timber." *American Journal of Agricultural Economics* 64 (2): 232–41.

Addicott, Stephen. 2019. "Stafford's Secondary Approach to Timberland Investment." Presentation at RISI Timberland Investment Conference, London, UK, May 8, 2019.

African Development Bank. 2017. *Unlocking Capital Flows for Forest Sector Development in Africa.* AfDB CIF Knowledge Series. July 2017.

———. 2019. *Towards Large-Scale Commercial Investment in African Forestry.* Study for Climate Investment Funds Evaluation and Learning Initiative.

Allison, David. 2015. "SEC Orders Atlanta's Timbervest to Pay $585,000." *Atlanta Business Chronicle*, September 22, 2015. https://www.bizjournals.com/atlanta/news/2015/09/22/sec-orders-atlantas-timbervest-to-pay-585-000.html.

Arano, Kathryn G., and Ian A. Munn. 2006. "Evaluating Forest Management Intensity: A Comparison among Major Forest Landowner Types." *Forest Policy and Economics* 9 (3): 237–48.

Aronow, Mary Ellen, Clark S. Binkley, and Courtland L. Washburn. 2004. "Explaining Timberland Values in the United States." *Journal of Forestry* 102 (December): 14–18.

Assogba, Noel Perceval, and Daowei Zhang. 2020. "Conservation Reserve Program and Timber Prices." Working paper, Auburn University School of Forestry and Wildlife Sciences.

Bailey, Conner, Abhimanyu Gopaul, Ryan Thomson, and Andrew Gunnoe. 2020. "Taking Goldschmidt to the Woods: Timberland Ownership and Quality of Life in Alabama." *Rural Sociology.* doi.org/10.1111/ruso.12344.

Bastin, J.-F., Y. Finegold, C. Garcia, D. Mollicone, M. Rezende, D. Routh, C. M. Zohner, and T. W. Crowther. 2019. "The Global Tree Restoration Potential." *Science* 365 (6448): 76–79.

Bearrows, James. 1986. "Other Financial Alternatives concerning Timberland Investment: Limited Partnership." In *Timberland Marketplace for Buyers, Sellers, and Investors and Their Advisors: Proceeding of 1984 Meetings*, 17–21. Durham, NC: Center for Forestry Investment, School of Forestry and Environmental Studies, Duke University.

Berman, Phillis. 1996. "Viva St. Spotted Owl!" *Forbes*, February 12, 1996, 44–45.

Berry, John M. 1984. "James River Builds Paper Empire." *Washington Post*, December 10, 1984.

Binkley, Clark S. 1994a. "Long-Run Timber Supply: Price Elasticity, Inventory Elasticity and the Capital-Output Ratio." *Natural Resource Modelling* 7:163–81.

———. 1994b. *The Russell-NCREIF Timberland Index: A Review.* Vancouver, Canada: Faculty of Forestry, University of British Columbia.

Binkley, Clark S., Fiona Stewart, and Samantha Power. 2021. *Pension-Fund Investment in Forestry.* Washington, DC: World Bank. https://openknowledge.worldbank.org/bitstream/handle/10986/35167/Pension-Fund-Investment-in-Forestry.pdf.

Borchersen-Kato, Sarah. 2017. "Timberland REIT Potlatch to Acquire Deltic Timber in All-Stock Deal." Nareit. https://www.reit.com/news/articles/timberland-reit-potlatch-acquire-deltic-timber-all-stock-deal.

Boyd, William. 2015. *The Slain Wood: Papermaking and Its Environmental Consequences in the American South.* Baltimore: Johns Hopkins University Press.

Bradley, William, H. 2013. "Federal Income Tax—100 Years of Shaping Timberland Transactions: A Review of Past and Current Tax Policy and Some Predictions for the Future." Paper presented at Who Owns America's Forests Conference, Portland, OR, September 17–19, 2013.

Burak, Steven G. 2001. "Selection of the Expected Rate of Return in the Valuation of Timberland: The Determinants of the Timberland Discount Rate." PhD diss., Auburn University.

Busch, J., J. Engelmann, S. C. Cook-Patton, B. W. Griscom, T. Kroeger, H. Possingham, and P. Shyamsundar. 2019. "Potential for Low-Cost Carbon Dioxide Removal through Tropical Reforestation." *Nature Climate Change* 9 (6): 463.

California Environmental Protection Agency Air Resources Board. 2015. *Compliance Offset Protocol: U.S. Forest Projects.* June 25, 2015. https://ww3.arb.ca.gov/cc/capandtrade/protocols/usforest/forestprotocol2015.pdf.

CalPERS. 1985. *Minutes of Investment Committee Meeting.* July 18, 1985.

———. 1987. *Annual Financial Report of 1987.*

———. 1988a. *Hancock Timber Investment Information to Members of the Investment Committee.* January 19, 1988. Sacramento, CA: CalPERS Investment Office.

———. 1988b. *Investment Committee Meeting Minutes: John Hancock Timberland Portfolio Review.* September 20, 1988.

Cambridge Associates. 2017. *Timber Impact Investment Benchmark Preview: A Study for Global Impact Investment Network.* Boston: Cambridge Associates.

Campbell, Duncan. 1986. "Foreign Capital in U.S. Timberland Investment: Who Invests, Why, and What the Futures May Bring." In *Timberland Marketplace for Buyers, Sellers, and Investors and Their Advisors: Proceedings of 1984 Meetings,* 38–42. Durham, NC: Center for Forestry Investment, School of Forestry and Environmental Studies, Duke University.

Campbell Global. 2012. "The Campbell Group Acquires South Australian Forestry Assets." October 17, 2012. https://www.campbellglobal.com/about/press-release?mediaID=M127.

———. 2018. "OneFortyOne Purchase of Nelson Forests Confirmed by Overseas Investment Office." August 20, 2018. https://www.campbellglobal.com/about/press-release?mediaID=M141.

The Campbell Group. 1984. *Business Plan Summary.* Portland, OR.

Canal Industries Inc. 1987. *Celebrating 50 Years of Excellence in the Forest Products Industry: 1937–1987.* Conway, SC: Canal Industries Inc.

Carlton, Dennis W. 1979. "Vertical Integration in Competitive Markets under Uncertainty." *Journal of Industrial Economics* 27 (3): 189–209.

Carthew, James. 2011. "Cambium Global Timberland IT: Why Money Doesn't Grow on Trees." Citywire Funds Insider, June 14, 2011. http://citywire.co.uk/money/cambium-global-timberland-it-why-money-doesn-t-grow-on-trees/a594817.

Chang, Sun Joseph. 1981. "Determination of the Optimal Growing Stock and Cutting Cycle for an Uneven-Aged Stand." *Forest Science* 27 (4): 739–44.

Chapman, C. M. 1967. "A Private Banker Looks at Timber Loans." *Forest Farmer* 26:16.

Chiriac, Marian. 2014. "Much of Romania's Forest Restitution Deemed Illegal." *BalkanInsight,* October 17, 2014. https://balkaninsight.com/2014/10/17/illegalities-hamper-romania-s-forest-restitution-process/.

Chung, Juliet, and Dawn Lim. 2017. "Harvard Endowment Chief Pushed for Steeper Devaluation of Assets." *Wall Street Journal,* December 15, 2017. https://www.wsj.com/articles/harvard-endowment-chief-pushed-for-steeper-devaluation-of-assets-1513252800.

Clarke, M. 1999. "Devolving Forest Ownership through Privatization in New Zealand." *Unasylva* 50 (199): 35–44. http://www.fao.org/docrep/x3030e/x3030e0a.htm.

Clemmer, Cindi. 1985. "Edwin Craig Wall Sr.: The Man behind the Empire." *Myrtle Beach Magazine,* Winter 1985, 52–57, 68.

Clephane, Thomas P., and Jeanne Carroll. 1982. *Timber Ownership, Valuation, and Consumption Analysis for 97 Forest Products, Paper, and Diversified Companies.* August 25, 1982. New York: Morgan Stanley.

Clutter, Mike, Brooks Mendell, David Newman, David Wear, and John Greis. 2005. *Strategic Factors Driving Timberland Ownership Changes in the U.S. South.* https://www.iatp.org/sites/default/files/181_2_78129.pdf.

Collins, Charles. 2001. "Examples of Recent Timberland Transactions." In *Industrial Timberland Divestitures and Investments: Opportunities and Challenges in Forestland Conservation,* edited by Nadine E. Block and V. Alaric Sample, 29–31. Washington, DC: Pinchot Institute.

Conroy, Robert, and Mike Miles. 1989. "Commercial Forestland in the Pension Portfolio: The Biological Beta." *Financial Analysts Journal* 45 (September/October): 46–54.

"Crown Pacific Declares Bankruptcy." 2013. *The Forestry Source,* August 2013, 6.

DANA Limited. 2016. *The New Zealand Forest Products Industry Database Review 2016.* 10th ed. Rotorua, New Zealand: DANA Limited.

Daniels, Barbara J., and William F. Hyde. 1986. "Estimation of Supply and Demand for North Carolina's Timber." *Forest Ecology and Management* 14:59–67.

Dasos. 2019. *Future Prospects for Forest Products and Timberland Investment*. 5th ed. September 2019. Helsinki, Finland: Dasos.

Davis, Gerald F. 1991. "Agents without Principles? The Spread of the Poison Pill through the Intercorporate Network." *Administrative Science Quarterly* 36:583–613.

Davis, Gerald F., and Tracy A. Thompson. 1994. "A Social Movement Perspective on Corporate Control." *Administrative Science Quarterly* 39:141–73.

de Lisser, Eleena. 1998. "Georgia-Pacific Plans New Class of Stock Tied to Its Profitable Timber Business." *Wall Street Journal*, September 18, 1998.

Dezember, Ryan. 2018. "Tree Gluts Uproot Southern Investors." *Wall Street Journal*, October 10, 2018.

Draffan, George. 1999. *The Global Timber Titans: Profile of Four U.S. Wood Products Corporations Driving the Globalization of the Industry*. June 1999. https://bdgrdemocracy.files.wordpress.com/2011/12/globaltimbertitans.pdf.

Drăgoi, Marian, and Veronica Toza. 2019. "Did Forestland Restitution Facilitate Institutional Amnesia? Some Evidence from Romanian Forest Policy." *Land* 8:99. doi:10.3390/land8060099.

Ellefson, Paul V., and Robert N. Stone. 1984. *U.S. Wood-Based Industry: Industrial Organization and Performance*. New York: Praeger.

Fairbairn, Madeline. 2020. *Fields of Gold: Financing the Global Land Rush*. Ithaca, NY: Cornell University Press.

Fallon, Ivan. 1991. *Billionaire: The Life and Times of Sir James Goldsmith*. Boston: Little, Brown.

Fama, Eugene F., and Kenneth R. French. 1993. "Common Risk Factors in the Returns on Stocks and Bonds." *Journal of Financial Economics* 33 (1): 3–56.

Faustmann, M. 1849. "On the Determination of Value Which Forest Land and Immature Stands Possess for Forestry." In Institute Paper 42, translated by M. Gane. Commonwealth Forestry Institute, Oxford University, 1968.

Federal Reserve Bank of St. Louis. 2009. "Market Yield on U.S. Treasury Securities 10-Year Constant Maturity." https://fred.stlouisfed.org.

Fletcher, G. A. 1967. "Timberlands as Long-Term Loan Investment for Insurance Companies." *Forest Farmer* 26:15.

Fu, Chung-Hong. 2013. *A Look at the New NCREIF Timberland Fund and Separate Account Index*. Timberland Investment Resources. https://1nzy1a2az6m43b6rbr2f9hib-wpengine.netdna-ssl.com/wp-content/uploads/2014/06/A-Look-at-the-NCREIF-Fund-Index-2013-07-11.pdf.

———. 2014. *Timberland Investments: A Primer*. Brookline, MA: Timberland Investment Resources.

Gilbert, Adrian M. 1984. "Investing in Timberlands—The Hutton Case." *The Consultant*, January 1984, 14–16.

Global Impact Investing Network. 2017. *GIIN Perspectives: Evidence on the Financial Performance of Impact Investments*. https://thegiin.org/assets/2017_GIIN_FinancialPerformanceImpactInvestments_Web.pdf.

Goldman Sachs. 1983. *Our Expanding Universe: The Case for Pension Fund Investment in Property*. New York: Goldman Sachs.

Griscom, B. W., J. Adams, P. W. Ellis, R. A. Houghton, G. Lomax, D. A. Miteva, W. H. Schlesinger, D. Shoch, J. V. Siikamäki, P. Smith, and P. Woodbury. 2017. "Natural Climate Solutions." *Proceedings of the National Academy of Sciences* 114 (44): 11645–50.

Guan, Hongshu, and Ian A. Munn. 2000. "Harvest Restrictions: An Analysis of New Capital Expenditures in the Pacific Northwest and South." *Journal of Forestry* 98 (4): 11–16.

Gunnoe, Andrew A. 2012. "Seeing the Forest from the Trees: Finance and Managerial Control in the U.S. Forest Products Industry, 1945–2008." PhD diss., University of Tennessee.

"Hancock Gets Clearance to Buy Carter Holt Harvey's Forestry Assets." 2006. *New Zealand Herald*, September 27, 2006. https://www.nzherald.co.nz/carter-holt-harvey/news/article.cfm?o_id=36&objectid=10403251.

Hancock Timber Resource Group. 1986. *A Timber Investment Program for the California Public Employees Retirement System (CalPERS): Proposal—Timberlands RFP #85-31*. Boston: Hancock Timber Resource Group.

———. 1996. *Research Notes: The NCREIF Timberland Index*. Boston: Hancock Timber Resource Group.

Hartman, Richard. 1976. "The Harvesting Decision When a Standing Forest Has Value." *Economic Inquiry* 14:52–68.

Haynes, Richard W. 1998. *Stumpage Prices, Volume Sold, and Volumes Harvested from the National Forests of the Pacific Northwest Region, 1984 to 1996*. General Technical Report PNW-GTR-423. Portland, OR: US Forest Service Pacific Northwest Research Station.

Holmes, John A., ed. 1998. *The Dictionary of Forestry*. Bethesda, MD: Society of American Foresters.

Hourdequin, Jim. 2013. "Some Thoughts on the Future of TIMOs and REITs." Paper presented at the State and Future of Forestry in the U.S. Meeting, May 29, 2013, Washington, DC.

Howard, Theodore E., and Susan E. Lacy. 1986. "Forestry Limited Partnerships: Will Tax Reform Eliminate Benefits for Investors?" *Journal of Forestry* 84 (12): 39–43.

Howell, Rosemary. 2008. *The Fountain Forestry: The First Fifty Years*. Ilfracombe, United Kingdom: Arthur H. Stockwell Ltd.

Hungerford, Norman D. 1968. "An Analysis of Forest Land Ownership Policies in the United States Pulp and Paper Industries." PhD diss., State University College of Forestry at Syracuse University.

Hyde, William F. 1980. *Timber Supply, Land Allocation and Economic Efficiency*. Baltimore: Johns Hopkins University Press.

International Finance Corporation. 2019. *Creating Impact: The Promise of Impact Investing*. Washington, DC. April 2019. https://www.ifc.org/wps/wcm/connect/66e30dce-0cdd-4490-93e4-d5f895c5e3fc/The-Promise-of-Impact-Investing.pdf?MOD=AJPERES.

"International Paper and Power." 1937. *Fortune* 16 (6): 229.

International Paper Company. 1948. *After Fifty Years*.

———. 1998. "Generation of Pride: A Centennial History of International Paper Company." Reprinted as "A Short History of International Paper: Generations of Pride," *Forest History Today*, 1998, 29–34.

International Woodland Company (IWC). *25 Year Jubilee: 1991–2016*. Frederiksberg C, Denmark.

Janzik, Robert M. 1986. "Role of Pension Funds in Timberland Investment: Overview of Our Approach." In *Timberland Marketplace for Buyers, Sellers, and Investors and Their Advisors: Proceedings of 1984 Meetings*, 111–13. Durham, NC: Center for Forestry Investment, School of Forestry and Environmental Studies, Duke University.

Jenson, Michael J. 1969. "Risk, the Pricing of Capital Assets and the Evaluation of Investment in Stock Portfolios." *Journal of Business* 42 (2): 167–247.

"Jimmy Goldsmith's U.S. Bonanza." 1983. *Fortune*, October 17, 1983, 125–36.

Kirk, Preston E. 1985. "A Tale of Tall Timber: Trying Times in the Tree Business Offer Investment Opportunity for Some." *Muse Air Monthly*, March 1985, 15–18.

Lehman Brothers. 2006. *Timber 101: Valuing Timberland*. New York: Lehman Brothers.

Lehtonen, Petri. 2019. *Forest Finance Hub Led by FAO? A Concept Note*. Rome: Food and Agriculture Organization of the United Nations.

Li, Yanshu, and Daowei Zhang. 2014. "Industrial Timberland Ownership and Financial Performance of U.S. Forest Products Firms." *Forest Science* 60 (3): 569–78.

Linklater, Andro. 2013. *Owning the Earth: The Transformation History of Land Ownership*. New York: Bloomsbury USA.

Lintner, John. 1965. "The Valuation of Risk Assets and the Selection of Risky Investments in Stock Portfolios and Capital Budgets." *Review of Economics and Statistics* 47:13–37.

Lönnstedt, Lars. 2007. "Industrial Timberland Ownership in the USA: Arguments Based on Case Studies." *Silva Fennica* 41 (2): 379–91.

Maggard, Adam, and Daowei Zhang. 2020. "Why Did Stumpage Prices Remain Low When Lumber Prices Soared to Historic Highs?" *Alabama TREASURE Forest Association Newsletter* 6 (3) (December 2020): 1.

Marino, Vivian. 2007. "For Some Investors, Money Grows on Trees." *New York Times*, May 27, 2007. http://www.nytimes.com/2007/05/27/realestate/commercial/27sqft.html?_r=0.

Mason, George. 1986. "Role of Insurance Companies in Timberland Investment." In *Timberland Marketplace for Buyers, Sellers, and Investors and Their Advisors: Proceedings of 1984 Meetings*, 163–67. Durham, NC: Center for Forestry Investment, School of Forestry and Environmental Studies, Duke University.

Mattey, Joe P. 1990. *The Timber Bubble That Burst: Government Policy and the Bailout of 1984*. New York: Oxford University Press.

McDonald, Michael, and Irina Vilcu. 2015. "Harvard Exits Romanian Timber Woes, Selling Forests to Ikea." Bloomberg, July 23, 2015. https://www.bloomberg.com/news/articles/2015-07-23/harvard-exits-romanian-timber-woes-selling-forests-to-ikea.

Mei, Bin. 2019. "Timberland Investments in the United States: A Review and Prospects." *Forest Policy and Economics* 109:1–7. doi.org/10.1016/j.forpol.2019.101998.

Mei, Bin, and Mike Clutter. 2010. "Evaluating the Financial Performance of Timberland Investments in the United States." *Forest Science* 56 (5): 421–28.

Mendell, Brooks. 2013. "Wood Supply Agreements, Part I: Pricing Mechanisms and Market Analyses." *Forisk Blog*, May 9, 2013. http://forisk.com/blog/2013/05/09/wood-supply-agreements-part-i-pricing-mechanisms-and-market-analyses/.

Mendell, Brooks, Neena Mishra, and Tymur Sydor. 2008. "Investor Responses to Timberlands Structured as Real Estate Investment Trusts." *Journal of Forestry* 106:277–80.

Mills, W. L., and William L. Hoover. 1982. "Investment in Forest Land: Aspects of Risk and Diversification." *Land Economics* 58 (February 1982): 33–51.

Mirova. 2019. *Land Degradation Neutrality (LDN) Fund: An Innovative Impact Investment Fund for Sustainable Land Use, with a Linked TA Facility.* Paris: Mirova.

Munn, Ian A., and W. Rhett Rogers. 2003. "Forest Management Intensity: A Comparison of Timber Investment Management Organizations and Industrial Landowners." *Southern Journal of Applied Forestry* 27 (2): 83–91.

Murray, Brian. 1995a. "Measuring Oligopsony Power with Shadow Prices: U.S. Market for Pulpwood and Sawlogs." *Review of Economics and Statistics* 77:486–98.

———. 1995b. "Oligopsony, Vertical Integration, and Output Substitution: Welfare Effects in the U.S. Pulpwood Markets." *Land Economics* 71 (2): 193–206.

Neilson, Dennis. 2006. "Timberlands Ownerships in an International Context." *Western Forester*, March/April 2006, 6–7.

Newman, David H. 1987. "An Econometric Analysis of the Southern Softwood Stumpage Market: 1950–1980." *Forest Science* 33 (4): 932–45.

Newman, David H., and David N. Wear. 1993. "Production Economics of Private Forestry: A Comparison of Industrial and Nonindustrial Forest Owners." *American Journal of Agricultural Economics* 75:674–84.

New Zealand Institute of Economic Research. 2000. *Devolving Forest Ownership through Privatisation: Processes, Issues and Outcomes.* Working Paper 2000/3. May 2000. Wellington, New Zealand.

Niquidet, Kurt, and Glen O'Kelly. 2009. "Forest-Mill Integration: A Transaction Cost Perspective." *Forest Policy and Economics* 12:207–12.

Nyland, Ralph D. 2007. *Silviculture: Concepts and Applications.* 2nd ed. Long Grove, IL: Waveland Press.

Oden, Jack P. 1973. "Development of the Southern Pulp and Paper Industry, 1900–1970." PhD diss., Mississippi State University.

O'Laughlin, Jay, and Paul V. Ellefson. 1982. "Strategies for Corporate Timberland Ownership and Management." *Journal of Forestry* 12:784–88.

Pearse, Peter H., Daowei Zhang, and Janette Leitch. 1995. "Trends in Foreign Investment in Canada's Forest Industry." *Canadian Business Economics* 3 (3): 54–68.

Perry, Martin K. 1989. "Vertical Integration: Determinants and Effects." In *Handbook of Industrial Organization*, edited by Richard Schmalensee and Robert Willig, 103–255. Amsterdam: North-Holland.

Pitas, Claudia Roberts. 1986. "Timberland Investment by Pension Funds." In *Timberland Marketplace for Buyers, Sellers, and Investors and Their Advisors: Proceedings of 1984 Meetings*, 176–78. Durham, NC: Center for Forestry Investment, School of Forestry and Environmental Studies, Duke University.

Plum Creek. 2014. *Gratitude, Stewardship, Opportunity: 25 Years of Plum Creek.* Seattle: Plum Creek Timber Co.

"A Poison Pill Has the Pros Choking." 1985. *Business Week*, February 11, 1985, 96.

Porter, Earl, and William Consoletti. 2015. *How Forestry Came to the Southeast: The Role of the Society of American Foresters.* Cenveo Publisher Services.

Proskauer. 2016. "Significant Changes to U.S. Taxation of REITs and Investments by Non-U.S. Investors in Real Property under the PATH Act." February 2, 2016, https://www.proskauer.com/alert/significant-changes-to-us-taxation-of-reits-and-investments-by-non-us-investors-in-real-property-under-the-path-act.

Ranaivoson, James, and Enrico Canu. 2012. "EIB's Approach towards Forestry Funds." Presentation at roundtable meeting, Plantation Forestry in Developing Countries: Enhancing Innovation, Tackling Complexity, Luxembourg, May 22, 2012.

Rayonier. 2016. "Rayonier Pacific Northwest Portfolio Repositioning Supplemental Materials." May 2016. https://ir.rayonier.com/static-files/055f4b41-445c-4de6-b5fa-7004c29c7a97/.

————. 2020. "Rayonier To Acquire Pope Resources." January 15, 2020. https://ir.rayonier.com/news-releases/news-release-details/rayonier-acquire-pope-resources/.

Redmond, Clair H., and Frederick W. Cubbage. 1998. "Risk and Returns from Timber Investments." *Land Economics* 64 (November 1998): 325–37.

Resources for the Future. 1958. *Forest Credits in the United States: A Survey of Needs and Facilities*. Washington, DC: Resources for the Future.

Reuters. 2019. "Finland's UPM to Invest $3 Billion to Uruguay." July 23, 2019. https://uk.reuters.com/article/uk-upm-kymmene-oyj-results-uruguay/finlands-upm-to-invest-3-billion-in-uruguay-idUKKCN1UI0UV.

Rinehart, James A. 1985. "Institutional Investment in U.S. Timberlands." *Forest Products Journal* 35 (5): 13–18.

————. 2010. *U.S. Timberland Post-recession: Is It the Same Asset?* R&A Investment Forestry, April 2010. Accessed December 31, 2016. http://investmentforestry.com/resources/1%20-%20Post-Recession%20Timberland.PDF.

Rinehart, James A., and Paul S. Saint-Pierre. 1991. *Timberland: An Industry, Investment, and Business Overview*. 2nd ed. Prepared by Pension Realty Advisors Inc. under a contract funded by the Hancock Timber Resource Group. Boston: Hancock Timber Resource Group.

Robinson, Vernon L. 1974. "An Econometric Model of Softwood Lumber and Stumpage Markets: 1947–1967." *Forest Science* 14:177–93.

Ruseva T., E. Marland, C. Szymanski, J. Hoyle, G. Marland, and T. Kowalczyk. 2017. "Additionality and Permanence Standards in California's Forest Offset Protocol: A Review of Project and Program Level Implications." *Journal of Environmental Management* 198:277–88.

Sharpe, William F. 1964. "Capital Asset Prices: A Theory of Market Equilibrium under Conditions of Risk." *Journal of Finance* 19:425–42.

Sizemore, William. 1986. "Valuation and Appraisal." In *Timberland Marketplace for Buyers, Sellers, and Investors and Their Advisors: Proceedings of 1984 Meetings*, 193–202. Durham, NC: Center for Forestry Investment, School of Forestry and Environmental Studies, Duke University.

Smith, R., and K. H. Bacon. 1983. "Boom in Tax Shelters Artificially Lifts Prices of Much Real Estate." *Wall Street Journal*, December 27, 1983.

Smith, Richard N. 1985. "A Proposal for CalPERS Separate Timber Account." Memorandum to Richard P. Troy, January 28, 1985.

————. 1986. "Role of Insurance Companies in Timberland Investment." In *Timberland Marketplace for Buyers, Sellers, and Investors and Their Advisors: Proceedings of 1984 Meetings*, 203–7. Durham, NC: Center for Forestry Investment, School of Forestry and Environmental Studies, Duke University.

Smith, W. Brad, Patrick D. Miles, John S. Vissage, and Scott A. Pugh. 2004. *Forest Resources of the United States, 2002*. General Technical Report NC-241. St. Paul, MN: US Forest Service, North Central Research Station.

Stanbury, William T. 2000. *Environmental Groups and the International Conflict over the Forests of British Columbia, 1990 to 2000*. Vancouver, Canada: SFU-UBC Centre for the Study of Government and Business.

Stein, Peter R. 2011. "Trends in Forestland Ownership and Conservation." *Forest History Today*, Spring/Fall 2011, 83–86.

Sun, Changyou, Mohammad M. Rahman, and Ian A. Munn. 2013. "Adjustment of Stock Price and Volatility to Changes in Industrial Timberland Ownership." *Forest Policy and Economics* 26:91–101.

Sun, Changyou, and Daowei Zhang. 2001. "Assessing the Financial Performance of Forestry-Related Investment Vehicles: Capital Asset Pricing Model vs. Arbitrage Pricing Theory." *American Journal of Agricultural Economics* 83 (3): 617–28.

Sun, Kevin. 2020. "Timber REITs Are Having a Mighty Moment." *The Real Deal*, August 20, 2020. https://therealdeal.com/2020/08/20/timber-reits-are-having-a-mighty-moment/.

Sun, Xing, and Daowei Zhang. 2011. "An Event Analysis of Industrial Timberland Sales on Shareholder Values of Major U.S. Forest Products Firms." *Forest Policy and Economics* 13 (5): 396–401.

Sun, Xing, Daowei Zhang, and Brett Butler. 2015. "Timberland Ownerships and Reforestation in the Southern United States." *Forest Science* 61 (2): 336–43.

Thompson, Terri. 1985. "Luring Pension Mangers into the Woods." *Business Week*, February 11, 1985.

Timber Growth Corporation. 1998. *Amendment No. 1 to Form S-11 Registration Statement under the Security Act of 1933.* April 29, 1998. Washington, DC: Securities and Exchange Commission.

Timberlink. 2021. "Timberland Assets under Management as of December 31, 2020." Atlanta, GA: Timberlink. www.timberlink.net.

UBS RII. 1996a. *Timberland Investment - RII/Weyerhaeuser World Timberfund Executive Summary.* November 1996.

———. 1996b. "UBS Announces Closing of the World's Largest Timberfund Joint Venture." http://www.prnewswire.com/news-releases/ubs-announces-closing-of-worlds-largest-timberfund-joint-venture-75387432.html.

Ulrich, Alice H. 1990. *U.S. Timber Production, Trade, Consumption, and Price Statistics, 1960–1988.* US Forest Service Miscellaneous Publication No. 1486.

US Forest Service. 2010. "FIA Data Mart, FIADB Version 4.0." http://apps.fs.fed.us/fiadb-downloads/datamart.html.

Vance, Rupert B. 1935. *Human Geography of the South: A Study in Regional Resources and Human Adequacy.* 2nd ed. Chapel Hill: University of North Carolina Press.

Vardaman, James M. 1989. *How to Make Money Growing Trees.* New York: John Wiley & Sons.

Wang, Linda. 2011. *Timber REITs and Taxation.* General Technical Report. US Forest Service. August 2011. http://www.fs.fed.us/spf/coop/library/timber_reits_report.pdf.

"Weyerhaeuser Co.: Concern Seeks to Build Fund to Buy Foreign Timberlands." 1995. *Wall Street Journal,* April 14, 1995.

White House. 2016. *United States Mid-Century Strategy for Deep Decarbonization.* Washington, DC: White House.

Wiegner, Kathleen K. 1982. "A Growing Investment?" *Forbes,* November 8, 1982, 52–53.

Wilke, John R. 2004. "How Driving Prices Lower Can Violate Antitrust Statutes: Monopsony Suits Mount as Companies Are Accused of Squeezing Suppliers." *Wall Street Journal,* January 27, 2004.

Williams, Michael. 1989. *Americans and Their Forests: A Historical Geography.* New York: Cambridge University Press.

Wong, Peggy A. 1982. *Forestry Fund Investment Study.* May 1982. Marketing Research Department, US National Bank of Oregon.

World Bank. 2004. *Report of the Forest Investment Forum: Investment Opportunities and Constraints.* Washington, DC: World Bank. http://documents.worldbank.org/curated/en/143121468135583949/Report-of-the-Forest-Investment-Forum-investment-opportunities-and-constraints.

Yin, Runsheng, Thomas G. Harris, and Bob Izlar. 2000. "Why Forest Products Companies May Need to Hold Timberland." *Forest Products Journal* 50 (9): 39–44.

Yin, Runsheng, and Bob Izlar. 2001. "Supply Contract and Portfolio Insurance." *Journal of Forestry* 99 (5): 39–44.

Zhang, Daowei. 2007. *The Softwood Lumber War: Politics, Economics, and the Long U.S.-Canada Trade Dispute.* Washington, DC: Resources for the Future Press.

———. 2019. "Costs of Delayed Reforestation and Failure to Reforest." *New Forests* 50:57–70. doi.org/10.1007/s11056-018-9676-y.

Zhang, Daowei, Brett Butler, and Rao Nagubadi. 2012. "Institutional Timberland Ownership in the U.S. South: Magnitude, Location, Dynamics, and Management." *Journal of Forestry* 110 (7): 355–61.

Zhang, Daowei, and Richard Hall. 2020. "Timberland Asset Pricing in the United States." *Journal of Forest Economics* 35 (1): 43–67. http://dx.doi.org/10.1561/112.00000448.

Zhang, Daowei, and Peter H. Pearse. 2011. *Forest Economics.* Vancouver, Canada: University of British Columbia Press.

Zhang, Daowei, and Shaun M. Tanger. 2017. "Is There a Connection between Campaign Contributions and Legislative Commitment? An Empirical Analysis on the Co-sponsorship Activity of the 2007 Tree Act." *Forest Policy and Economics* 85 (1): 85–94.

Zinkhan, F. Christian. 1988. "The Stock Market's Reaction to Timberland Ownership Restructuring Announcements: A Note." *Forest Science* 34:815–19.

———. 1993. "Timberland Investment Management Organizations and Other Participants in Forest Asset Markets: A Survey." *Southern Journal of Applied Forestry* 17 (1): 32–38.

———. 2008. "History of Modern Timberland Investments." In *Proceedings of Conference on International Forest Investment (London, UK),* 1–19. Athens, GA: Center for Forest Business.

Zinkhan, F. Christian, William R. Sizemore, George H. Mason, and Thomas J. Ebner. 1992. *Timberland Investments.* Portland, OR: Timber Press.

Index